Team Working for Better Math Students

Team Working for Better Math Students

A Collaborative Approach for Parents and Teachers of Students in Grades 1-12

with interesting bits of math history as to why and how we learn Math plus Everyday Decision Making!

James Elander

Copyright © 2017 by James Elander.

ISBN: Softcover   978-1-5245-4178-1
             eBook   978-1-5245-4177-4

All rights reserved. No part of this book may be reproduced or transmitted in any form or by any means, electronic or mechanical, including photocopying, recording, or by any information storage and retrieval system, without permission in writing from the copyright owner.

Any people depicted in stock imagery provided by Thinkstock are models, and such images are being used for illustrative purposes only.
Certain stock imagery © Thinkstock.

Print information available on the last page.

Rev. date: 08/22/2017

To order additional copies of this book, contact:
Xlibris
1-888-795-4274
www.Xlibris.com
Orders@Xlibris.com
746818

Team Working for Better Math Students

A Collaborative Approach for Parents and Teachers of Students in Grades 1-12

with
short interesting bits of math history
as to why and how we learn Math
plus
Everyday Decision Making!

Prologue

This book is for the benefit of students in grades 1-12. That is quite an undertaking, but the objective is to provide explanations of the material so the parents can help their student. The content not only provides a background as to learning material, but also some pertinent history to make the subject more meaningful.

The author calls this a Teamwork Approach. Teachers and administrators that the author asked to reviewed the material were all in flavor of this approach.

Why the grades 1 to 12?

These are key years where many students, due to their expanding world, will be left behind if they are not encouraged and helped with their learning, especially in grades 6, 7, and 8. This book will aid the teachers and parents by providing bits of history as to the development of mathematics resulting in more meaningful and interesting math achievement. It will assist the parents to understand what is being taught and consequently they can help their student understand. It has been stated in many ways that you can tell the extent of nation's civilization by the extent of the mathematics obtained by the average citizen. As a civilization becomes more sophisticated, especially in a democracy, the people must become more educated in the area of decision making. The demands today require a life time of learning, which is always changing and is more demanding. Look at the number of new applications we have been challenged to learn and use in the last 30 years.

3
(That makes for an informative discussion with your student.)

Schools have made a mistake to not offer special programs to keep parents informed, especially in the early years as to the changing methods and content, so they can help their children succeed. The parents should become almost a teacher assistant, and will need some background in math.

The concepts are introduced in a historical, meaningful explanatory way that will assist the parents and teachers to be more helpful for their very important objective, making learning more interesting. The author realizes that parents may feel frightened by this approach and would be wrong if he stated it would not require some extra work.

An interesting addition in this book is the Bibliography of source books for teachers, parents and many for students to make the course more interesting plus providing help for the teachers, parents, and the students. You will be proud of the end result. Teachers and parents will be surprised at the enjoyment in reading some of these books. (You probably don't believe it, but try it. A good start for parents would be Abbott's FLATLAND written in the 1880s or THE EDUCATION OF T. C. MITS. You will be surprised!)

The author firmly believes in parents helping their children with their homework under the direction of the teacher as to the amount of homework and methods.

The history and background for the concepts and development of problems types are partly explained in this book. This is because parents many times are asked for help and explanations by their student. The book explains the Why and How plus some development. This approach will provide the answers to Why and How plus some interesting bits of history to make it more interesting.

Somebody once said:

Don't tell your children what to think, but teach them how to think!

Valid Decision Making is a major objective of mathematics and the basics are incorporated into this book. It should be in all math textbooks! You will enjoy this material! A democracy must have a well-educated critical thinking voting populace. Socrates and Plato started this approach to educate the public.

Comment

The suggested grade levels for introducing and teaching the concepts and skills vary due to the variation in the school programs and text books.

About the author

Jim Elander, retired former math teacher from positions in both high and college, also past President of two professional organizations, MMC and SSMA, a General Electric Fellow at Purdue University, and Math Department Chairman at Oak Park and River Forest High School in Oak Park, IL. He is also an author, and was an Assist. Professor at North Central College, Naperville, IL. Now retired in beautiful Missoula, MT.

He has other books on the market. (All his books have explanations for why and how with everyday applications for aiding future Decision Making Skills.)

TGIF MATH
A book consisting of over a hundred activities to make the hectic days, like prior to vacation, into exciting learning days. The students will look forward to these days. Available at your local bookstore or from the publisher: 888-793-4274 ext. 7879.

BASIC HIGH SCHOOL MATH REVIEW
A book to prepare students for the college or Tech School entrance exams with explanations as to why for better understanding, plus basic Decision Making Skills.
Prerequisites: Algebra and Geometry.
Available at your local bookstore or from the publisher: 888-793-4274 ext. 7879.

His first Geometry text was published in 1992 by South Western, since then he has written a much improved geometry text available on CDs or Flash Drives from the author. Each student has their own flash drive. (This results in a very small cost to schools instead of a costly set of hardbacks.)

    Jim Elander   (elanderje@gmail.com)
    Missoula, MT

A bit older now, but a better golfer!

Table of Contents

Prologue 2
About the author 5
Selected Bibliography 10
Chapter 1. The Early Years - Grades 1-3 15
   Section 1a. The first Challenge: Counting 16
   Section 1b. The Basic Operations and some History .. 21
   Section 1c. Some Early Geometry History 25
Chapter 2. The Foundation years - Grades 3-5 32
   Section 2a. The Golden Age of Mathematics 32
   Section 2b. The Order of Operations and Why 34
   Section 2c. Operations: Fractions and More History .. 36
   Section 2d. Some Related History 39
   Section 2e. Summary of Chapter 2 40
Chapter 3. The Middle School Prep Years - Grades 5-7 47
   Section 3a. The Golden Years of Greek Mathematics. 47
   Section 3b. Money and Percent 50
   Section 3c. Pictures of Mathematics 53
   Section 3d. Rounding numbers and the Effects 56
   Section 3e. The Game of Equations (An Introduction) 58
   Section 3f. Negative Numbers 61
Chapter 4. The High School Prep Years - Grades 6-9 65
   Section 4a. Algebra Formulas and Equations 65
   Section 4b. Direct Variation 67
   Section 4c. Graphs: Line graphs, $y = Mx + B$ 71
   Section 4d. DESCARTES' GIFT 75
   Section 4e. The Addition Method 80
   Section 4f. A New Kind of Number 83
   Section 4g. The Substitution Method 86
   Section 4h. Graphing Short Cuts 90
Chapter 5. Introduction to Statistics - Grades 9-10 95

Section 5a. Averages 95
Section 5b. A bit of the History of Statistics 105
Chapter 6. High School - Grades 9-10 108
Section 6a. Why Study Mathematics? 108
Section 6b. What is Mathematics? 110
Section 6c. Why is Geometry different? 113
Chapter 7 International System of Measurement 123
Section 7a. The World's Measuring Units 123
Section 7b. SI Conversion Table 125
Chapter 8. High School Years – Grades 9-12 128
Section 8a. Plane and Solid Geometry Review 128
Section 8b. Algebra Summary 135
Section 8c. Graphing Inequalities and Applications .. 144
Section 8d. Exponents 150
Section 8e. Logarithms 155
Section 8f. Trigonometry for Right Triangles 166
Section 8g. Trig. for Non-Right Triangles 179
Section 8h. THE NORMAL CURVE 190
Section 8i. Types of Statistical Graphs 197
Section 8j. Casino Math Probability 201
Section 8k. Odds, Expectation, and Fair Bets 207
Chapter 9. Logical Decision Making - Grades 9-12 216
Section 9a. Use and Misuse of Statistics 216
Section 9b. Inductive Conclusions 226
Section 9c. Illusions, Games and Applications 231
Section 9d. Forms of an Implication and their uses .. 239
Section 9e. Basics for Decision Making 243
Chapter 10. The Student's Future 246
Comments for the Parents and Teacher 246
Chapter 11. Course Review – Grades 10-12 250
Section 11a. Geometry 250
Section 11b. Algebra Review 258

Section 11c. Decision Making Review 265
Index 1. Basic Postulates 277
Index 2. Definitions 279
Index 3. Essential Geometry Theorems 286
Index 4. Quotes 294
Index 5. Conversion Table 304

Selected Bibliography

A listing of interesting, informative and unique books that parents and teachers will enjoy, plus many for your students. Give them a try!

(Check your school or public libraries, they may have them. **Books marked with an * are suggested especially for students and/or parents and teachers.**)

    Abbott, Edwin A.
        FLATLAND A ROMANCE OF MANY DIMENSIONS (very **interesting)***
        Princeton University Press

    Banks, Robert B.
        SLICING PIZZAS, RACING TURTLES, AND FURTHER ADVANTURES IN APPLIED MATHEMATICS*
        Princeton University Press

    Beckmann, P.
        HISTORY OF PI*
        St. Martin's Press

    Bell, E. T.
        MEN OF MATHEMATICS*
        Simon & Schuster

    Byrkit, D.
        "TAXICAB GEOMETRY."*
        MATHEMATICS TEACHER, May 1971, Pages 418-422

    Cajori, Florian
        HISTORY OF ELEMENTARY MATHEMATICS
        The Macmillan Company

    Davis, J.J.
        BIBLICAL NUMEROLOGY*
        Baker Book House (Pi value is stated in the Bible, erroneously, I Kings 7:23)

Davis, P. and Hersh, R.
THE MATHEMATICAL EXPERIENCE*
Houghton Mifflin

DESCARTES DREAM*
Harcourt Brace Javanovich

Devlin, K
Mathematics-the new golden age
Columbia University Press

Dudley, Underwood
NUNEROLOGY or, What Pythagoras Wrought*
Mathematical Association of America

MATHEMATICAL CRANKS*
Mathematical Association of America

Eves, H.
Great moments in Math before 1650

Fadiman, Clifton
THE MATHEMATICAL MAGPIE*
(Mobius Strip-"Paul Bunyan vs. The Conveyor Belt)"
Simon and Schuster

Fawcett, Harold
NATURE OF PROOF* (Teacher, Parent)
13th Yearbook of NCTM

Florman, S. C
ENGINEERING AND THE LIBERAL ARTS HS,* (parents and teachers)
McGraw-Hill Company
(A guide to History, Literature, Philosophy, Art, Science, and Music)

Gardner, M
MATHEMATICAL CARNIVAL*
Alfred A. Knopf Publisher

MATHEMATICAL CIRCUS*
Vintage Books
Division of Random House

Gazale, Midhat
   NUMBER: From Ahmes to Cantor
   Princeton University Press

Gordon, Sheldon and Florence, Editors
   STATISTICS FOR THE TWENTY-FIRST CENTURY
   Mathematical Association of America, 1992

Hogben, L, Sir
   Mathematics For The Millions*
   The Wonderful World of Mathematics
   Oxford

Huff, Darrell
   HOW TO LIE WITH STATISTICS*
   Norton Co.

Katz, Victor **A**
   HISTORY OF MATHEMATICS
   Harper Collins

Kenny
   "Hemholtz And The Nature Of Geometric Axioms"
   Mathematics Teacher, Vol. 50, Feb. 1957

Klein H. A.
   THE WORLD OF MEASUREMENTS*
   Simon and Schuster

Kline, M.
   MATHEMATICAL THOUGHT FROM ANCIENT TO MODERN TIMES
   Oxford University Press

Lieber, L.
   MITS, WITS, AND LOGIC*
   Institute Press, New York, 1954

   THE EDUCATION OF T. C. MITS* (Who is T.C. Mits?)
   W. W. Norton & Co., 1954

13

Loomis, E.
>**THE PYTHAGOREAN PROPOSITION.**
>NCTM publication
>**Comment**: (Which former President of the U.S. is credited with a proof?)

Nolan, Deborah, Editor
>**WOMEN IN MATHEMTICS: SCALING THE HEIGHTS***
>Mathematical Association of America

Northrop, E. P.
>**RIDDLES IN MATHEMATICS*** (A Book of Paradoxes)
>Van Nostrand Company

Packel, Edward
>**THE MATHAMATICS OF GAMES AND GAMBLING***
>Mathematical Association of America

Paulos, J.
>**I THINK, THEREFORE I LAUGH***
>Vintage Books
>Division of Random House

Peterson, I.
>**THE MATHEMATICAL TOURIST***
>W. H. Freeman and Company

Poe, Edgar Allen
>**THE GOLD BUG*** (A Mystery involving mathematical reasoning.)

Polya, G.
>**MATHEMATICAL DISCOVERY: Vol. 2** (Chapter 14: The art of teaching mathematics.)
>John Wiley & Sons

Postman, N.
>**TECHNOPOLY***
>Alfred A. Knopf

Reid, Constance
>**A LONG WAY FROM EUCLID***
>Thomas Y. Crowell Co.

Reeve, W. D.
> **THE TEACHING OF GEOMETRY** (For the teacher)
> 5th Yearbook, NCTM

Stevenson, R. L.
> **TREASURE ISLAND* (chapter 31 or 34)**
> (Locus problem-location of the treasure.

Weber, R.
> **A RANDOM WALK IN SCIENCE***
> Crane, Russak & Co. Inc.
> "Life on Earth.(by a Martian")
> (Fascinating little story (p. 124) with a surprise ending.

Video or film (Grades 7-12)
> **DONALD DUCK IN MATHMAGIC LAND* Disney (good for a PTA meeting) also)**

An interesting critical thinking test.
> Critical Thinking Test, Level X by R. Ennis and J. Millman
> (Very interesting and a different type of test based on a space travel theme. I have given this to several hundred high school and college students on a pre/post test basis and to my surprise the below average group gained the most.
> Available at Foundation for Critical Thinking, 1-800-833-3645 or 1-800-458-4849

Web sites:
> **www.criticalthinking.org**
> archives.math.utk.edu
> www-history.mcs.st-and.ac.uk
> www.MAA.org
> www.nsf.gov
> **www.AMS.org**
> mathforum.org

Add your own selections

Chapter 1. The Early Years - Grades 1-3
A bit of Math History

MATHEMATICS has been a human activity for thousands of years (and) to some extent everybody is a mathematician and does MATHEMATICS.

P. Davis and R. Hersh

The Mathematical Experience

We often read in the paper, hear on radio or on TV, comments that people make relative to student progress and/or discipline. Your parents, or grandparents, may remember this first line of an old song.

Reading, writing and 'rithmetic, taught to the tune of a hickory stick.

The following is from a Sumerian composition (about 2000 B.C.) relating to education called *SCHOOL DAYS. (E. Kramer's THE NATURE AND GROWTH OF MODERN MATHEMATICS. Page 1.)*

I went to school...I read my tablet, ate lunch...prepared my tablet... went home...my father was proud of me...I said wake me early in the morning for I must not be late or my teacher will cane me."

Have times changed much? (What is a tablet, your student may ask?)

The early years, but really at any age, the topic of counting is an interesting and major objective. Students of all ages are interested in this evolution and it is natural to weave it into the learning process. The major TV networks have missed an opportunity, in my opinion, by emphasizing the negative experiences instead of the positive ones. Hollywood has also failed in the area of cinema emphasis, especially for families. Reminder: This material is written for the parents and teacher. It should be read by them first, so they will be able to help the student interests with some historical facts, but use the teaching method the teacher is using.

Chapter 1. The Early Years - Grades 1-3
Section 1a. The first Challenge: Counting

There cannot be a language (Mathematics) more universal ... and more worthy... to express relations
Joseph Fourier

This material is for grades 1-3. Be sure to check with the teacher as to the content to be taught, when it will be covered and the methods used plus the expectations.

How did the method of counting evolve? This is a good topic for discussion and research. When do you think it started and how? Let's go back to 2000-3000 years B.C. and think of ourselves as a farmer trying to make a living for his family. Fortunately, you have a small herd of cattle or sheep, and every day you let the herd out to eat and hope they will all come back in the evening. (Why didn't the author suggest times like 5 or 6 pm?) As time goes by the herd gets larger and you need a method of determining that the animals are all back. What could they do? (Listen to your student's suggestions. You may even want to record the various suggestions.) Remember they had no counting system!

Eventually the students will arrive at what we call comparing sets. In other words when an animal leaves, a pebble could be put in the bag and when it returns in the evening you take a pebble out. If the bag is empty when they are all "home," and you know the complete set of animals is back. (I used the word SET since it isn't really counting, but does indicate quantity due to the one to one correspondence. The topic of Set Theory wasn't really formulated until the 1800s A.D. by Georg Cantor (1845-1918) and introduced into the schools about 1960.) If you are more interested check out a math history book from your library. (See the bibliography for suggestions. You will be able to refer to it many times in this program.

The actual counting systems (notice plurality) evolved as the demand from the business and the construction worlds required the need to know how many. A logical system is the base 10 system corresponding to relationship between the ten fingers and ten objects. (The base 10 system was also used by some American Indians. Cajorie, p.5.) The number 1 would be illustrated by 1 finger, 2 fingers → the number 2, 3 fingers for 3, and etc. in the Roman Numeral system of counting illustrated this by I = 1, II = 2, III = 3, IIII = 4, V = 5. Where did the V come from? If you hold your hand closed in front of your face (say at an arm's length) and indicate in order a 1, then a 2, and so forth for 3, 4, and 5. Doesn't the hand representing 5 resemble a V? Problem solvers are always looking for easier ways to do things and the Roman Numeral symbol for 5 became a V. Can you now see why and how the X became the symbol for 10. (Do you see the two Vs?) This was an easier way to write ten in Roman Numerals!

There is another system, base 2, which is very useful today. It is related to the computer and electricity, on-off (base 2). There were many numeral systems invented to indicate the numbers, The Roman Numeral system is just one that has survived. (The word numeral refers to the symbols that represents the numbers.)

The two terms, number and numeral are used interchangeably. All ancient civilizations had created their own systems. (This makes for some interesting research as to the method your ancestors learned.) Another system which we use is the one on the face of a clock, 1-12. Military time is another type, more on this later.

The Babylonians had quite a sophisticated system (3000-2000 BC). A system of numerals used for weights and measure in construction and a monetary system used in banking for simple and compound interest. The base was 60 instead of 10. (Page 3 in Kramer's *THE NATURE AND GROWTH OF MODERN MATHEMATICS)*

The drawback to all the early systems was that they did not have a symbol for zero and they had to invent symbols for each number. A few additional examples using Roman Numerals are: C for 100, L for 50, M for 1000, these came from the words for centurion (century) and millennium (1000). I think the symbol for 50 may have evolved from the lower half of the numeral C. Writing the lower half of the C carelessly several times and it approaches an L. (This is the author's theory for the evolution of the symbol L for 50?) In one system, the symbol π represented 9,000,000, I think it was the Babylonians (F. Cajorie) HISTORY OF ELEMENTARY MATHEMATICS, 1896). Our system is base 10, but another base could have been base 20. Why? (counting fingers and toes.) Which would mean how many additional symbols would be needed? Another interesting point is that the symbol 0, which was not invented early A.D. (Research: Which group of people invented the symbol for the number 0? It was the Hindus.) The logic up to then was, why have a symbol for nothing. You can see the great advantage using the 0 enables us to write a numeral for any size number (using only the ten digits), instead of having a unique symbol for every number. Examples: 01, 10, 100, 101, 1000 a good way to explain this and help for understanding is to use a model of a car odometer. Start with an odometer with no decimal point. And start indicating the counting as shown on the odometer. (Good activity for parents & student at home work on.)

$$\boxed{0\ 0\ 0\ 0}$$

I also found the following method in F. Cajorie's book this interesting method for adding and multiplying numbers.

To add and multiple arrange the problem the following way. This method could possibly help some students. (Notice it goes from right to left. This was a Hindu method.) Very interesting!

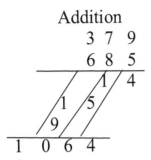
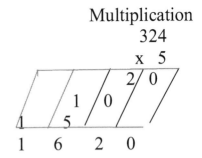

An interesting quote:

The genesis of knowledge (the product of an education) must follow the same course as the genesis of knowledge of the forefathers.

F. Cajorie
HISTORY OF ELEMENTARY MATHEMATICS

I would add, but we must cover the material at a must faster rate in order to cover the many more advancements. Home activities should include cases of counting, measuring, and writing numerals. It is advisable to use objects from inside or outside when first teaching counting. (Always check with the teacher and have a common understanding.)

Using the concept of the odometer is an easy way to lead into carrying over to the next higher counting set and to introduce fractions. Using the following and even the one in the car can be very helpful. It takes time, but it is worth the time. (Notice, no decimal point at this stage!)

100s	10s	1s

The odometer helps in the reading of whole numbers and decimals. Grades 1&2, will complete the 1s and 10s and grade 3 will begin to understanding for the 100s and may even the begin the case for the decimals and/or fractions. The Roman B.C. period boys (not girls, why?) learned fractions (called broken numbers) using money, weights and measures. Some of these take time and the parents should reinforce the method the teacher is using. The parent's advantage is they have one on one situation and time.

The teacher may also introduce the class to simple geometric figures like the triangle, circle and the square. There are usually plenty of these figures in the house for the student to point out.(Check with the teacher.)

Notes and Comments

Chapter 1. The Early Years - Grades 1-3
Section 1b. The Basic Operations and some History

MATHEMATICS IS NOT A CAREFUL MARCH DOWN A WELL CLEARED HIGHTWAY, BUT A JOURNEY INTO A STRANGE WILDERNESS.

W. Anglin

This "highway" contains addition, multiplication, subtraction, division, estimation, measure, area and perimeter.

The next big step in your student's education is how to use these numerals in what we call the elementary arithmetic operations and what the answer represents. In other words, learning how to add, subtract, multiply and divide. Adding and subtracting are basic operations since Multiplication is repeated Addition and Division is repeated Subtraction. This may sound odd, but if the problem is to multiply 3 times 2 it can be solved by adding 2+2+2 = 6 or (2 x 3 = 6). This also shows that addition and multiplication are **commutative**, or A x B is the same as B x A. If the problem is to divide 10 by 5, then 10-5-5 = 0 and the student can conclude that 10 divided by 5 is 2. Another way is to use the multiplication tables. Example: 5 times 2 is 10, or 10 -5-5 divided by 2 is 5, from the table. Is division commutative (is A divided by B the same as B divided by A)?

Your great grandparents probably had to learn the times table up to 15, your parents up to 12, and some schools today require the table memorized up to10. This will be explained later on. What is the teacher's objective? The parents really need to work with the student to learn and use these basic operations. (Use physical objects when applicable.) Question: Is subtraction commutative?

Your great grandparents were possibly taught a method for checking the answers for basic operation problems called **Casting Out Nines**. (Personally, I think it should still be taught in the schools today.) Here is an example using Casting Out Nines to check an addition problem. The problem is to add the following.

Adding the digits:

$$\begin{array}{r} 123 \rightarrow 6 \rightarrow 6 \\ 456 \rightarrow 15 \rightarrow 6 \\ + \; 987 \rightarrow 24 \rightarrow 6 \\ \hline 1566 \rightarrow 18 \rightarrow 9 \end{array} \bigg\} \rightarrow 18 \rightarrow 9$$

The check is to add the digits in 1566 (your answer) which is18 and adding again gives 9. The two sums are the same are the same. Your answer is correct.

Now add the digits in the numbers you are adding. (sums shown above.) and both answers are 9, therefore the answer 1566 is correct. Do you see why it is called "Casting Out Nines"? This method works for the 4 operations. If the problem is subtraction, then in the check you subtract, if multiplication, then you multiply, if division, then check by your answer times the divisor.

Check these answers by Casting Out Nines:
a. $135 + 791 + 246 = 862$ b. $94 \times 32 = 3008$
c. $108 \div 3 = 36$ d. $92 - 21 = 71$

Check for c: 108 should equal 36x3 and it does 9=9
e. Another sample: $12436 \times 973 = 12100228$
Is the answer correct? Check by Casting Out Nines.
Yes: The digits of the product and factors are each 7.

Create a few more problems over x days for each operation and check the answers by Casting Out Nines. (Some of your friends may fine this method interesting! There is one defect in the method. Do you see it? (Hint: order of the digits.)

```
           1237        7321       7643      567 ÷ 9 = 63?
         + 598       - 945     x  809
Answers:   1835        6576     6183187
```

Be sure to understand the method the teacher is teaching and use it along with the above over x days if you wish. The (B.C.) Hindu teachers used the above method for checking answers.

It is very gratifying to observe your student is understanding when the operations involve fractions and decimals. Usually teachers start with unit fractions (numerator is 1). (1/2, 1/3, 1/4 and reinforce the understanding by using objects) The learning of the HOW for the basic operations is a must for the early grades and something the parents should assist their children to make them feel successful and develop an "I can do it attitude."

The 4 basic operations are a must and the teacher should inform the parents as to the method being taught. It can be very confusing to the student if the same
methods are not used.

The following was a Russian method for teaching multiplication.

Objective is to find the product of 5 x 23. (It is assumed the student can add.)

$$\begin{array}{cc} 1 & 23 \\ 2 & 46 \\ 3 & 69 \end{array}$$

Now 2+3 is 5 so 46 + 69 is 115, the product.
The answer is 5 x 23 = 46 +69 = 115

Notes and comments

Chapter 1. The Early Years - Grades 1-3
Section 1c. Some Early Geometry History

It is not How Much You Cover, but How Much You Uncover.
H. Fawcett

Note: Harold Fawcett, author of the 13th yearbook of the NCTM, gave me advice at a meeting in Salt Lake City while we were walking around the square.

Check with the teacher as to when and what geometry will be introduced. The concepts of measure, estimation, perimeter, and area, plus names of geometric figures are easily learned and correlated with numbers. Again, know what the teacher at each grade level expects and the methods used for teaching it.

Here is an interesting figure which means the early mathematicians (1500 B.C.E.+) had an insight to the most important theorem and how they used the Pythagorean Theorem in Plane Geometry ($A^2 + B^2 = C^2$). This reads: In right triangle ABC, the square on side AC equals the sum of the squares on sides AB and BC. This is more meaningful to actually make the figure and justify it by cutting it out and showing the relationship. Use the **isosceles right triangle** below! The reason the credit for the theorem is given to Pythagoras is that he proved it for all cases. The following figure shows that the square on side AB plus the square on side BC equals the square on side AC.

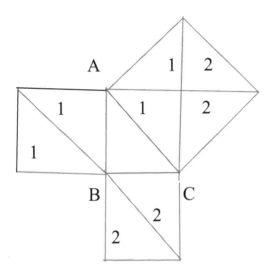

The large square is equal to the sum of the 2 small squares. This is a good activity for cut outs (Isoceles right triangle).

The statement is the square upon the hypotenuse (don't use the name) of the isosceles right triangle (ABC) is equal the sum of the two squares upon the equal sides of the triangle. It took the Greeks over a 1000 year later to prove the general case for all right triangles. The general case will be justified in the high school section of this book.

The years prior to 1000 B.C.E. were what I will call the Arithmetic years. These were years of applications in the trades, or learning by doing. An example is the problem pertaining to area. They knew the formula for the area of a square and probably the area of rectangle, but there was nothing in regard to a proof or justification for a formula. Their geometry could be called intuitive geometry.

Area example is:
   One Square unit is: therefore this figure has an area of 4 Sq units

It was observed by the early users that the area was 2x2 or 4 square units, which lead to the formula A = LW.

What are the perimeters of the above two figures? What is the definition of Perimeter and the formula? P = 4s

The area of a parallelogram (in bold) may have been known from the following:

   This ▱ be can be converted to this square ☐ by relocating the triangle to form the square.

This illustrates a basic problemsolving method, convert the problem to a type you know how to solve!

The following is a possible explanation as to how they developed the approximation for the area of a circle and the circumference. The question as to the area and the circumference of a circle are two problems which were finally settled in the 1870s. (See the book by Beckmann listed in the bibliography on Pi, especially the bit on Indiana.)

To the credit of the early mathematicians they did have approximations for the formulas related to a circle. Their method was to start with the same method as we use today based on what we know. They would like to be able to calculate the circumference or area of a circle given the radius. For the circumference they marked a point on the circle and rolled the circle until the point on the circle had completed one complete rotation and then measured the distance from A to A which is the measure of the circumference. The formula is C = K times R, R is the radius and K is the constant. Example: If C =30 units and R is 5, then 30 = 5K and therefore K equals 6. Naturally the value of K may come out a fraction depending on the accuracy of your measurements. (Use the wheel of bike)

A   A

This is a fun team learning problem for the parent and student to take the time and work it out. (It involves measurement, estimation, using possibly a bike, dividing to solve for K.) Try it using different size circles. Then look up the value or even the history of the value for this K which you know as 2π. (C = 2πR)

A possible method solution for the area of a circle:

For the area of a circle They could have started with the following figure and counted the squares then like in the case above calculate the k value in the equation A = kr^2. This activity will take considerable help from the parent, but will be worthwhile. The estimations should be a good learning exercise for your student. Take your time and help the student. (Probably grades 4-6).

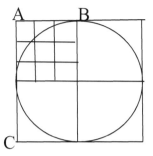

The area of the small square with side AB is 1/4 the area of the large square and the area of the circle is less than the area of the large square with side AC. Now help the student to estimate the area of the ¼ circle that is in the small square. Hint: 4+.9 + .9 +1.1 is the author's estimate for the area of the circle in ¼ of the circle. This sum times 4 is the author's estimate for the area of the circle or 27.6. The area of the large square is 36 square units. (The ancient mathematicians did not know the area of a circle is πr^2, but they wanted to know was what number times the radius squared gives the area. Therefore: from our estimate we can write 27.6= k(9) or k is 3.1 rounded off. The symbol, π, was not assigned for the constant until the 1700s by L. Euler.

Another early solution method for the Area of a circle:
For the Area of a circle they started with the formula for a square based on the length of a side, but they wanted the area of a circle based on the radius. They discovered that a regular hexagon with a side of R could be inscribed in a circle. They also knew the area of the circle was greater than the area of the hexagon (6 equilateral triangles) and less than the area of the large square or 4 small squares.

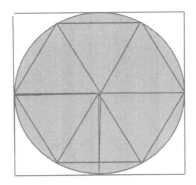

Comment: More on this topic in the high school section.

This created some interesting conclusions. Theologians concluded that the earth rotated in 360 days, which became the 360 degrees for a complete rotation of a circle(6x60). This also led to sixty degrees in the arc that is determined by the side of the hexagon. The Babylonians, 2000 BC, realized the area (in the above figure) of the circle was greater than the area of the hexagon and less than the area of the circumscribed square. They had arrived at values for Pi (it was not labeled Pi until after 1700 A.D. by Euler.) at 3 and 1/8, or in decimal form 3.14. It is also in the Bible (1st Kings 7:23). What is the Biblical value of Pi?) Interestingly, the question of Pi was not settled until 1872 A.D. by Lindemann. This could be classified as the problem that took the longest time to solve in the history of math. (over 4000 years) Chinese had also calculated a value for Pi. Naturally the base 60 was used to arrive at 360 degrees in a circle, 60 minutes in a degree and 60 seconds in a minute. The symbols came later. (Teacher and/or parents may enjoy the book *1421* (that is the title) with regard to the Chinese history.)

31

Your student's teacher may have introduced the names of some Geometric figures, a few like triangle or rectangle and their properties. Look at the following picture and ask what the student sees. You may need to give some hints. (A parent may wish to create a slide show of geometric figures in the buildings in the community and entertain the class. The teacher would appreciate it also. Your student will be proud of you.)

Chapter 2. The Foundation years - Grades 3-5
Section 2a. The Golden Age of Mathematics (1000 B.C. to 100 A.D.)

MATHEMATICS IS THE GATE AND THE KEY TO ALL SCIENCES, HE WHO IS IGNORANT OF IT CANNOT KNOW THE THINGS OF THIS WORLD.
ROGER BACON

Arithmetic, intuitive Geometry, and some Algebra from 3000 BC-1000 BC saw that Mathematics had advanced to meet the needs of the society, but the Greek period is labeled The Golden Age of Mathematics, 1000 B.C.– 100 A.D. This time period provided a period that had special advances in all categories due to societal mental enrichment. Even today people travel to visit these historic sites to recollect and observe the evidence with amazement of that era. But the non-physical and visual one we will visit is the one in the World of Mathematics. This period was the first development of rigor or the justification of necessary conclusions. (Justification by Rigor means that the conclusion is based on undefined terms, defined terms, assumptions, and theorems or laws that are previously justified by the first three.)

Benjamin Pierce, in the late 1800s A.D., defined math this way:

Mathematics is the science of drawing necessary conclusions.

This definition is a very meaningful application of the definition is Jefferson's **Declaration of Independence.** To observe this just read the first few pages of the Declaration. Note the definition does not mention or restrict mathematics to numbers, but all logical conclusions, valid or invalid. It was mentioned a few pages back that Math is taught for two objectives, one is its everyday application in areas involving number, finances, navigation, etc. The other objective is drawing valid conclusions or decision making. This objective is for all individuals to understand, especially in a democracy. We will have much more on this later and in theory this is an objective of every high school education. Plato had at the entrance to his school:

LET NO MAN ENTER HERE IGNORANT OF GEOMETRY.

PLATO (ca. 400BCE)

A practical application used by the surveyors in B.C.E. along the Nile River to determine the boundaries of lots. They had what we would call a tool made of rope, marked off as shown in the following figure. They were called rope stretchers, we would call them surveyors.

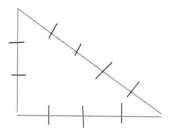

(Project: Make the 3-4-5 tool out of string and observe how it works to determine right angles.)

Do you recognize any relationship? It is a good activity to actually make this rope "tool" and stretch it out and let the student observe the result. More on this topic later on! The next chapter will also relate to one of their problems.

Chapter 2. The Foundation Years - Grades 3-5
Section 2b. The Order of Operations and Why

IN THE ACIENT WORLD AS NOW, TRADE HAS BEEN THE PRINCIPLE CONSUMER OF MATHEMATIC OPERATIONS.

P. DAVIS AND R. HERSH

The first case we will visit is a simple one, but is important in order to move forward to others cases. It is how to evaluate a numerical expression. Now as Algebra became more useful and necessary, the evaluating a numerical expression that time about 1000 B.C demanded the order and, yes, they even had an income tax. The following is an example and why.

Example: $10 - 6 \times 2 + 2\ 4 \div 8 + 7 = ?$
What is your answer, or answers?

The explanation is in the method for counting money? We will use the simple case of determining the amount of money in what is called the piggy bank.

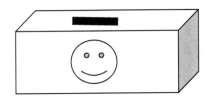

Did you record your answer?

You are probably thinking, why is this important? I assume you may, or a group of you, had several answers to the above. The order demanded by the banks is easily understood if you empty your pocket or the piggy bank and try to count the change. Try it! The procedure will show you why! (Incidentally, the correct answer to the above evaluation is 8.)

Here is the explanation:

What you do is first is separate the coins into groups. Now assume this is the contents of your piggy bank.

So, you have: Pennies, nickels, dimes, quarters, and half dollars as indicated below.

| | 15 | 20 | 18 | 9 | 2 |

Evaluate by multiplying by their values:

Value: $.15 $1.00 $1.80 $2.25 $1

Now add for the total: Answer is $6.20.

What you did to find the value was first multiply and then add to get the sub totals $15(.01) + 20(.05) + 18((.10) + 9(.25) + 2(1)$!

The answer depends on the order in which numerical operations are simplified. This was settled by the bankers of Egypt and is called the Order of Operation, by this is meant the order in which these operations are used to simplify a series of operations. You may have recalled it by the phrase some of you had to memorize as My Dear Aunt Sally method. Notice the order of the first letter in each word MDAS. Many teachers taught this to remind their students to first multiply, then divide followed by addition and subtraction. THE PROBLEM WITH THIS MDAS approach is that it doesn't answer the question, Why? (You probably recall when your children were 3 to 7 years old and were always asking WHY, until they finally gave up and just took the parent's word.) People of all ages should continue to ask WHY on all issues. Try a couple more problems (below) to insure your understanding. Einstein advised us to never stop asking WHY!

    a. $X = 18 - 2 \times 3 + 30 \div 5 - 3 \times 7 = ?$

    b. $X = 32 + 4 \times 6 - 8 \times 7 + 10$

        Answers: -3, 10

Chapter 2. The Foundation Years - Grades 3-5
Section 2c. Operations: Fractions and More History

GOD GAVE US THE INTEGERS AND ALL THE REST IS THE WORK OF MAN (AND WOMAN).

L. KRONECKER

Here is another problem that needs explaining for the teacher and parents.

    a. $24 \div 1/2 = ?$ Try another one. b. $1/2 \div 2/3$

What is your answer for a? 48 or 12? For b? 2/6 or ¾ ?
        Correct answers are 48 and ¾

Division is the same as multiplication and subtraction is the same as addition but each is the inverse of each operation. In the 1960s (I think) there was a show on TV (Art Linkletter was the M.C. called "Out of the mouths of kids" or something similar to that.) During one of the shows he asked big kids the following type of question. What is $8 \div 1/2 = ?$ One person came back 3 times with 3 different answers!

What is your answer?

Can you prove or justify your answers to a/b or a(1/b)?

The general case is $N \div (a/b)$. What is your answer? How was it taught to you? This is why it works, or the proof for it. (It is actually multiplied by 1.)

$$N/(a/b) \text{ or } \frac{N}{(a/b)} = \frac{Nb}{(ab/b)} = \frac{Nb}{a}$$

DO YOU SEE THE 1? (b/b)

Which shows the rule (when you divide by a fraction -invert and multiply) which you were taught and memorized! Were your answers to a and b above correct?

Operations with fractions are more difficult for beginning students. Read carefully how their text and the teacher approaches the topic and the parents should follow the same method. Probably start with unit fractions like ½, 1/3, ¼, etc. (use objects to help your student understand fractions). The figure below represents the rectangle divided in half. This shows that half of the total rectangle contain represents an X or represents 1/2. (Check with the teacher for their suggestions and the method used.)

Students need a lot of practice to understand and master the operations with fractions, and **parents need to have a lot of patience.** The operations of adding and subtracting fractions with different denominators is more difficult for beginners. Using objects is recommended.   1/2   1/2
  a. Draw a figure to show that ½ x ½ is ¼.
       ½ of (1/2) = 1/4
  b. Draw a figure to show that ½ + ½ is 1.
  c. Draw a figure to show that 1/3 + 1/3 is 2/3.
  d. Draw a figure to show that 1/2 ÷ 1/3 is 3/2.
  e. Draw a figure to show that 1/2 x 1/3 is 3/2.
  f. Draw a figure to show that 1/2 + 1/3 is 5/6.
       Hint: 3/6 + 2/6 = 5/6

Or ½ is 3/6 and 1/3 is 2/6, and 3/6 + 2/6 is 5/6, hence the answer is 5/6, as shown above.
(Emphasize: Convert each to the same units)

The above is difficult for beginners and is the reason decimals, which are much easier to use and are used internationally. Also, the example above, shows why we need common denominators in order to add fractions.

Parents will need to help the student with various types of examples and have much patience. It should be interesting for the parents to observe their student to learn to add and multiply fractions! Be sure to work with the teacher and use the same methods.

Add your notes or observations

Chapter 2. The Foundation Year - Grades 3-5
Section 2d. Some Related History
Number Rules The Universe

Pythagoreans

ALEXANDER THE GREAT ASKED HIS TEACHER IF THERE WAS AN EASIER WAY TO LEARN GEOMETRY? HIS TEACHER REPLIED: "THERE IS NO ROYAL ROAD TO GEOMETRY"

(I would change to "all mathematics.")

F. CAJORI

The Golden Age (1000 B.C.E. to 0) was quite different from the preceding 2000 years, since it was unique in many areas, art, literature, construction, music, government, navigation, chemistry, science, and of course the Queen of the Sciences, especially in Geometry. In this time period, the leaders such as Aristotle, Plato, Pythagoras, Euclid, Eratosthenes, Theon and Hypatia, Thales, and others also had academies. (Search each name using your computer, one person's will surprise you! A woman - why?) These leaders knew that a democracy requires a group of well-educated leaders, hence schools were started for mainly the upper class. (Research Plato, Euclid and Zeno, First and Second Alexandrian, the Sophists, the Pythagoreans.) Business men, like Thales, who traveled and encountered the business world of many countries as far East as China and brought back the positive customs and advancements from these countries. He and other business man observed and gained an understanding of these applications that could benefit their homeland. More History in the next chapter.

Mathematics is taught for two major objectives, which was mentioned prior. The objective of this little book is to be useful and helpful, for the parents with some related math history so the parents can help their student in grades 1-12, plus make the courses more enjoyable. The two major reasons are: (1) the practical, everyday math, and (2) the critical thinking.

Now some homework for parents and student! Be patient with your student! Homework is essential, just like practice is in a sport. The school's text should have the homework as assigned by the teacher. If you have questions check with the teacher.

Chapter 2. The Foundation Years - Grades 3-5
Section 2e. Summary of Chapter 2

(Note: Be selective with help of the teacher.)

Students... the first time something new is studied they seem hopelessly confused... then upon returning to the concept or method ...everything has fallen in place.
E.T. Bell

The following are just a few interesting activities to basically help your student to understand the home work exercises. (You may have to be selective as to which of these problems are related to what the teacher's assigned as homework and taught in preparation for the homework.)

41

(It is wise to look over the material before you try help and assist the problems with your student. Your student may not be ready yet for some of these.)

1. Count the numerals in this set.

1, 1, 3, 3, 4, 2, 5, 6, 7, 4, 8, 9, 6, 5, 3, 2, 1

2. How many 1s? 2s? 3s? 4s? 5s? 6s? 7s ? 8s? 9s? are there in problem 1?
3. What is the sum of the numbers the numerals represent?
4. Which numerals in #1 use only straight line segments to write.
5. Which numerals in #1 use circles or arcs?
6. Measure, using a ruler, the length of each line segment to the nearest centimeter, then to the nearest inch, then to the nearest of a millimeter.

a. A ―――――――――――――――― B

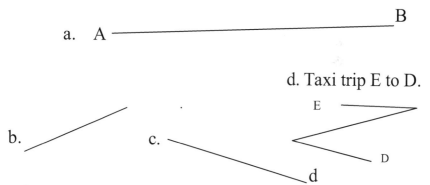

b.

c.

d. Taxi trip E to D.

Question: How can the length of an arc or part of a circle, be measured?

Note: You may ask the student to estimate the lengths of some objects in the house using feet, inches, or yards. (metric units if they have studied them.)

7. What is the sum of the numerals in #1 that use only line segments? (Example: the numeral 4 uses only 4 segments.)
8. What is half of the numeral 8? Answer is 0 or 3.
9. What is half of the number 8? Answer is 4.
10. $10 + 4 \times 2 - 7 + 2/(1/3) - 5 = ?$ Answer is 12.
11. Ask your student what is the product when multiplying by 1?
12. Draw a table like in #13 and complete it for A + B. Is A+B the same as B+A?
13. Complete the following multiplication table with your student for A times B Table. (Complete another table for B times A.)

B \ A	1	2	3	4	5	6	7	8	9
1	1	2	3	4	5	6	7	8	9
2	2	4							
3	3								
4	4								
5	5								
6	6								
7	7								
8	8								
9	9								

14. Ask your student: What is the answer when multiplying by 0?
15. Draw a table like in #13 and complete it for A - B.
16. Is A-B the same as B-A?
17. Someday the student will ask; What do I do when the numbers get larger?

You may find the following useful to make multiplication make easier.
   e.g. 7 x 15 is?   Ask is 15 = to 10+5?
The student will agree therefore the easy way is 7(10 + 5) =?
Then 7 x (10 + 5) is 7x10 + 7x5 or 70 + 35 and they can add this, hence 7 x 15 is 105.

18. If a number is divisible by 2. Then what does the number end with, even or an odd number?
19. Write the first 15 counting numbers. List the ones which are divisible by 2? by 3, by 5?
20. Pythagoras is given credit for forming a mathematics club which was devoted to the development of mathematics (more on this club later). One of their studies was figurative numbers. Which numbers could be represented by geometric figures? Try equilateral Triangles.

Name and draw the figure for a few more.
   If n is 4, figurative name is square.
   If n is 5, figurative name is pentagon.
   If the side is 6, figurative name is hexagon.

21. The fraction is 1/7, what does this mean?
22. Ask the student to write the first nine counting numbers and circle the one that is written the poorest, then perform the following. Multiply the circled digit by 9 and then multiply the product by 12345679. The answer will surprise your student and possibly you. Parent should try this first!
23. a. ½ + ½ =?   b. ½ times ½ =?   c. ½ - ½ =?
    d. ½ ÷ ½ =?   e. 2/3+1/2 =?   f. 2/3 - ½ =?

The answers to problems 9, 12. 16, 20, 23, 24 are listed below.
Answers to 9. yes, 12. Yes, 16. No.
The next two triangle figures are 12 and 15.
The next two Square figures are 9 and 16.
Answers to 23:
   a. 1   b. ¼   c. 0   d. 1   e. 7/6   f. 1/6

Comments: Did the teacher assigned homework to cover any geometric figures? If so then practice drawing them (use a ruler), naming them and counting the segments or sides and points or corners. If three dimensional figures were studied, then point these out in the house, community or in pictures in magazines.

Extra activities
1.

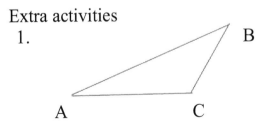

Points 3, Segments 3. Measure to the nearest Millimeter: AB=?   BC=?   AC=?

a. Complete the following: **Replace** in the following the ? marks with < for (less than) or > for (greater than) or = for (equal to) so that the statements are true and complete:
AB + BC ? AC
AC + BC ? AB
AC + AB ? BC

b. Write a general conclusion relating the answers.

c. Do the same as for these:
AB − AC ? BC
AB - BC ? AC

2a. What does your student see in the following, a cube in the corner or a cube outside the big cube? May have to blink a few times! Students and adults seem to enjoy pictures like these.

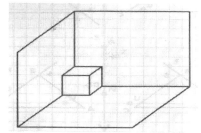

b. What does your student see in the picture below? (Look from the right and then from the left.)

(This is why people can report differently in jury cases.)

Notes and comments

Chapter 3. The Middle School Prep Years - Grades 5-7
Section 3a. The Golden Years of Greek Mathematics

MATHEMATICS SEEMS TO ENDOW ONE WITH SOMETHING LIKE A NEW SENSE.

CHARLES DARWIN

The period between 1000 B.C.E. to 800 C.E. Some called them the Golden Years of Mathematics, since it set some methods and trends that even exist even today, especially in decision making. Forms of government were also investigated and possibly tried, resulting in the democracy form of government. Their greatest contribution in mathematics was in Geometry, some label it as contributions undreamed of. The abacus played the role that a calculator plays today.

Suggested computer research: Thales (600 B.C.E.), Plato (450 B.C.E.), Euclid (300 B.C.E.), Pythagoras (550 B.C.E.), Archimedes (250 B.C.E.), Eratosthenes, Theom, Hypatia, Zeno (450 B.C.E.) and others.

Some of the interesting contributions are:
   a. Plato's Academy, which was an adult (men only) school that prepared men for logical thinking and leadership roles.
   b. Euclid's (300 B.C.E.) geometry text written in a logical deductive manner emphasizing not only conclusion pertaining to geometric figures, but what all decisions are based on. (Logical decisions are based on 4 items, undefined terms, defined terms, assumptions, and previous decisions.) A perfect example of the above in Jefferson's Declaration of

Independence. Read the first few pages with your student and observe how Jefferson arrived at the conclusion.

Euclid's Geometry consisted of 13 books (chapters) and is even a guide for texts today, even at the high school level. In the 1800s, it was a college course at Oxford and other colleges. This was probably a result of Plato's requirement at his adult academy and certainly Socrates Academy. He recognized it as a model course for teaching Critical Thinking or Decision Making (See the bibliography). Many of our geometry courses today neglect this important objective and don't emphasize it.

Another person you will find interesting is Pythagoras (550 B.C.E.), who had formed a secret Math Club for math enthusiasts. He is known for what is classified as the most important theorem in basic geometry, The Pythagorean Theorem. (In a right triangle, $a^2 + b^2 = c^2$ where c is the side opposite the right angle.) One story is that a member revealed secret and suddenly disappeared! The story is that a member revealed that the square root of 2 is irrational. This theorem will be proved in the high school section of this book.

There are at least currently two "secret" membership math clubs today. Their club names are Bourbaki (after a former mathematician in Europe) and Nancago (named after the universities at Nancy and Chicago.)

Another interesting scientist-mathematician is Archimedes (250 B.CE.). One story is how he defeated the Roman Army with his inventions (Many more interesting readings listed in the Bibliography).

Eratosthenes (250 B.C.E.) measured the circumference of the earth. How did he do it? (check the math history books or do a computer search)

India and China had developed many of the same geometric facts, especially in navigation. (See the book titled *1421.*) A few Math History books discuss their contributions. Some were similar to the Greek contributions and some more advanced. (Math concepts are the same in all civilizations. Number systems may be different.)

These grades (1-6) are the ones that put the finishing touches on the basics and creates the foundation for success in the new material that really begins in grades 9 to 12. The term basics may have different definitions depending on your school and in recent years even in classes. As far back as in the 1940s, larger schools, population wise and wealthy communities, had accelerated classes. The demands for science ability students resulted in the Advanced Placement programs in the post-World War II period. In this chapter, the following will be introduced:

> Metric system, Rounding off, negative numbers, Simple equations, tables and graphs, square roots, some geometry, and per cents.

Check with the teacher as to the topics and other topics studied. Always use the method used by the teacher and/or the textbook! (Check the books listed in the Bibliography for interesting related readings.)

Comment for teacher: The author used to bring in math books, many that are listed in the Bibliography, to high school classes on the day before a long vacation period and ask the students to select a book and start reading it. This was instead of the ordinary class period. They were happy. Then near the end of the period they were informed, if they wished, they could take the book home during vacation. It was amazing how many books were checked out and the opinions of mathematics changed!

As I mentioned before, a great advancement in counting was the "invention" of zero around 700 A.D. by the Hindus. This simplified the number system, since it permitted the writing of any number using only the 10 digits and the decimal system. (Some of these sections may not be covered by the student's text book and/or the teacher. Check with the teacher for the topics and methods!)

Chapter 3. Middle School Prep Years - Grades 5-7
Section 3b. Money and Percent

HISTORY SHOWS THAT THOSE HEADS OF EMPIRES WHO HAVE ENCOURAGED THE CULTIVATION OF MATHEMATICS...ARE ALSO THOSE WHOSE REIGNS HAVE BEEN THE MOST BRILLIANT AND WHOSE GLORY IS THE MOST DURABLE.

Michel Charles

Students usually are interested in studies related to money! Wonder why?

Even during the years prior to 1000 B.C.E. the financial world was very active and was concerned about interest rates and the mathematics involved. Business men even were using Simple and Compound Interest. And the first symbols for percent (meaning 1/100 is recorded around 1500 +/- A.D. Some of the symbols for percent they used and introduced when translated were pc, 2pc, p,c, 2p/c, o/o, no/o and finally n% or 3%. The printing press eventually began to have an influence on the symbols. Therefore, 5% now means 5/100 or .05 rather than fractions. Decimal fractions were invented by Simon Stevin about the late 1500s, but using the decimal point was invented by John Napier (1600s), probably the demand by the business and science areas required a more common unified system. Decimal numbers are much easier to work with in evaluations then working with fractions. Example: Which is easier to add .25 + .20 or ¼ + 1/5? Percent is mostly associated with financial problems of money. Students usually enjoy these problems. We will close this section with the Banker's Rule of 72, which many teachers and parents don't understand. Its application has caused many problems on long periods of time. Before we close this section, it should be mentioned that originally the business world used negative numbers to represent debt and the science world to represent distances, but a symbol for zero was not needed, they felt.

Teacher and parent must work together to avoid student misunderstandings and lack of success. You may wish to introduce simple and compound interest definitions and concepts. You may even wish to start a savings account for the student. Be sure to explain and answer any their questions, in other words make it a learning experience.

The **Banker's Rule of 72** states: Seventy-two divided by the annual rate of interest tells you the number of years it will takes for the money to double!

Example: You deposit $1000 in the bank at 9% at your child's birth and agree you will not withdraw the money until the student graduates from college or his 24th birthday.
The result using the rule of 72 is: (72/9 = 8)
Year: 0   8   16   24
Amount: $1000 $2000 $4000 $8000

Assume the student graduates with a good position and decides to leave the money in the bank until he or she is 64.

The result is:
Year: 32   40   48   56   64
Amount: $16,000 $32,000 $64,000 $128,000 $256,000

Do you think your student will be impressed? (A very negative example is credit card debt that uses months instead years!)

The students should practice converting simple per cent problems (3% to .03 and its meaning or use like 3% sales tax). These activities or homework may provide an opportunity to use the calculator, but always ask: Does the answer make sense?

There will be much more pertaining to this financial topic in the high school chapters. Let the student estimate the tips.

Chapter 3. Middle School Prep Years - Grades 5-7
Section 3c. Pictures of Mathematics

Understanding involves work, appreciation from applications.

Unknown

A Bit of History

Humans have always had a desire to travel and see what is over the hill, or beyond the horizon. The need for maps, which could be called pictures, became necessary to preserve routes for future business sources and explorers. The easy way is to follow the shore line, but this is not the fastest or shortest. One inventor who is given credit for today's maps is Mercator in the 1500s. (Computer search for the interested parent or teacher.) The source for our needs at this time is a road map for the student to read and determine distances. (For more information pertaining to map making do a computer search for H2G2 A BRIEF HISTORY OF MAP MAKING.)

The reading of graphs can easily be done from those in the paper or magazines. The construction of a graph can be done with help from the parent using data like the temperatures for a week, class grades on a test. Graphs are an easy short and colorful way of telling past history and many times providing a prediction for the future. Keep in mind the future can't be predicted with absolute certainty.

Each graph should have a title, and labels for the vertical and horizontal axes, be easy to read and tell the story, plus inform the reader where and how the data came about. The following bar graph tells a story about a high school sports program. Can you interpret it? You might make a similar one for the boys or girls in your school.

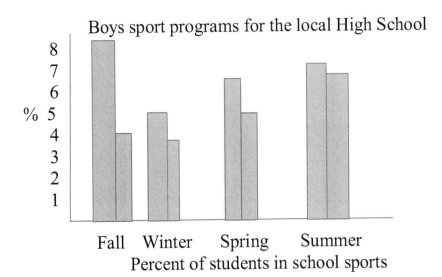

Questions:
1. Name the possible sports for each season.
2. How many are estimated to be in each sport for each season, if the school population is 500?
3. What are the participation numbers for the girls sports programs in your school? Make a graph.
4. What would you add to the graph to make it more informative?

Parent and Student Team Work Activity

55

Construct a line graph or one with related data objects for the following data. Help the student with a title, the data, the message and construction.

Subject data: What employers felt were important for employing high school graduates. (75 local employers were interviewed.)

Recommendations (by former 80% of the employers):

    Communications Skills: 75%
    Math Skills: 65%
    Business Skills: 55%
    Liberal Arts: 45%

Students find the circle graph the hardest.
Help your student made a circle graph to show the message the employers are stating? What is the message? Use the face of round clock face as a model. (The digits on the clock are 30° a part so estimating the angle size is easy.) Be sure to include the features listed below for the complete picture. Help your student to make a compete, accurate picture graph.
        Title
        Display
        The message or story
        Data for correct interpretation
Some graphs may tell a false or incomplete story since some people may see things differently as to how the data may be interpreted. Cut out a few graphs from a newspaper or a magazine and discuss them with your student.

CHAPTER 3. Middle School Years - Grades 5-7
Section 3d. Rounding numbers and the Effects

NEGLECT OF MATHEMATICS WORKS INJURY TO ALL KNOWLEDGE

Roger Bacon

There are several methods for rounding a set of numbers to the nearest integer.
    3.45, 2.56, 7.89, 9.12, 2.49.

Method 1: Add all the given numbers and then round for to the nearest integer.
    Answer is 25.51 then round to 26.

Method 2: Round the numbers in the set to the nearest whole number and then add.
    3 3 8 9 2
    Answer is 25

Method 3: We will add two numbers to the original set to illustrate this method.
    3.45, 2.56, 7.89, 9.12, 2.49, 5.57, 5.52.

Notice the new ones both could be rounded to 6, but this new method is to look at the last digit and if it is even then leave it as (5.52 as 5.5) as 5.5, but if the last digit is odd (5.57), then round it to 5.6 (5.57 as 5.6). The set then is 3.5, 2.5, 7.9, 9.1, 2.5, 5.6, 5.5. The rounded sum is 36.6, where the original sum is 36.6. Banks may use this method for rounding of number.

Comment: With a large set of data, method 3 seems to be the most valid. The reasoning is that there just as many even as odds so the + or − cancel each other. Use the method the teacher and your text are using, if this is part of the program. The rounded answers should always fit the data based on how it was obtained. Example: The average miles per gallon for a trip would not be calculated at 35.5678 miles per gallon, when you only used whole numbers to estimate the answer. The answer should not be more accurate than the numbers in the data set. Again, the answer should be sensible, it would be meaningless to give an area of 15 square feet as 2160 sq. inches. All measurements are approximate and are rounded off.

The business world many times works with negative numbers. These Numbers are indicated several ways by a − sign, by red ink, by a symbol like 3 in the hole.

$3

The key board or printer keys have set the method that is used today! When was the printing press invented? Even today some countries use ⌐ for the decimal point.

Until about 700 A.D. The symbol for zero was not really accepted. Why have a symbol for nothing? (The Hindu-Arabic system is basically what we use today. It was at first not use by some groups due to religious affiliations.)

<.... − 4 − 3 − 2 − 1 0 +1 + 2 + 3 + 4....>

Answers for addition or subtraction were determined by what could be called common sense. Like you have $5 and you owe $7 means – $2 in debt. Gas per gallon is price at say $3.259, then what is the cost of 10,000 gallons using $3.26 and then $3.259? Large quantities require rounding off methods for totals (like a bank with thousands of accounts). Ask your bank how they round off.

Research: How did Eratosthenes measure the circumference of the earth? See L. Hogben's book. It could be in your public library.

Chapter 3. Middle School Prep Years - Grades 5-7
Section 3e. The Game of Equations (An Introduction)

THINKERS MANY TIMES REALIZE THAT EQUATIONS PROVIDE THE ANSWERS TO A VERBAL QUESTION.

Unknown

Recently some programs, such as the Common Core, are introducing solution to equations for all students in the early grades. Where in former years, it was only for the special groups. It is a valid approach to introduce equations to all, but it is harder for the teacher and sometimes boring for better students.

First, what is an equation?

Definition: An equation is a statement that two numbers are equal.

Example: x + 5 = 12 → x = 7 (I am using → to mean therefore x is 7).

The following is not an equation: 3 feet equals 1 yard (3 ≠ 1). The statement is true, but it is not an equation. Note: ≠ means "not equal."

In order to play, the "game" of equations, you must know the rules, just like in any game. The rules for the game of equations are:
1. You may add or subtract the same number to each side of the equation.
2. You may multiply each side of the equation by the same number.
3. You may divide each side of the equation by the same number, except by 0.
4. The Order of Operations is used for simplification.

The equations in the text, which are assigned by the teacher, should be worked or solved the way the teacher teaches the students, even if the parents prefer another method.

Another example: Given: $5(3x + 2) = 6x - 8$
    Step 1: $15x + 10 = 6x - 8$ Why or How?
    Step 2: $9x = -18$ Why or How?
    Step 3: $x = -2$ Why or How?
    General Rule: **Try to reduce the problem to a simpler one**. Notice in the example each step after #1 is simpler.

Comment: Somebody once said: we learn the new from the knowledge of the past and others add, we don't learn from the past.

Equations were originally solved mostly by guessing or by trial and error and the result then provides a hint for a better guess. The early mathematicians called Rule 1: Inversion and Rule 2: Restoration (Parents were possibly taught it as Transposing).

The guess method: $3x+10=40$ What is the value of x?
Try x = 12 you then have 46 = 40 which is false, so try a lower value like 9.
    $29 + 10 = 40$ → $30 = 40$ which is false
So, try a higher value x=10
    $30 + 10 = 40$ → $40=40$ → correct so x=10.

   The rule method is faster.
      $3x + 10 = 40$
      $3x + 10 - 10 = 40 - 10$
      $3x = 30$
      $3x/3 = 30/3$
      $x = 10$

We shouldn't underestimate the ancient mathematicians, here is an example from F. Cajorie's, HISTORY OF ELEMENTARY MATHEMATICS, page 23.

AHMES' Papyrus about 2000 B.C.E. has this problem:
    $x(2/3+1/2+1/7+1) = 33$, what is the value for x?

Now help your student with the problems the teacher assigned.

Chapter 3. Middle School Prep Years - Grades 5-7
Section 3f. Negative Numbers

Why does a negative number times a negative number equal a positive number?

MATHEMATICS CONSISTS OF ISLANDS OF KNOWLEDGE IN A SEA OF IGNORANCE.
                                                        Simon Singh

(What does this quote mean?)

History: Negative numbers were use prior to 1000 B.C.E. as previously pointed out, but mostly related to an expression signifying debt. The Greek period or the Golden Age for Mathematics mostly developed the Geometry and related it to logic. Not much was improved in the subject of Algebra. The world of numbers was enlarged, by introducing the irrational numbers along with proof. More on this later.

At some point in the program the teacher will introduce negative numbers along with the operations using these new numbers. The student's number world at this time consists of only rational numbers. Rational numbers are one that can be expressed as a ratio of two counting numbers, like 2/3.

There is one justification the teacher should have. It is a proof that -1 (-1) = +1. Many teachers just tell the class it is a rule without the justification.

Here is one justification:
$-1(-1) = ?$, but we know that 1 times N (any number not 0) is N. Also, we know $1(-1) = -1$ and that $(-1)+1 = 0$, and 0 times any number is 0.

Now follow this argument for $-1(-1)=1$.
$-1\{1 + (-1)\} = -1(0) = 0$
But by the distributive property
this is $(-1)(1 + (-1)(-1)] = 0$
or $\quad -1 + (-1)(-1) = 0$.
Therefore: $(-1)(-1)$ must equal $+1$! Q.E.D.
What does QED mean? See a good dictionary.

Now use the homework the teacher assigns to practice adding, subtracting, multiplying, and dividing with negative numbers.

Suggestion: Try to make learning fun, but be patient with your student, and always use practical cases!

A Reading Problem

Question: How many were going to St. Ives?
Read this 18th century English nursery rhyme (which is a version of an Egyptian puzzle) very carefully, and then answer the question.

> *As I was going to St. Ives*
> *I met a man with seven wives.*
> *Each wife had seven sacks,*
> *Each sack had seven cats.*

> *Each cat had seven kits*
> *Kits, cats, sacks, wives,*
> *How many were going to St. Ives?*

Answer: Read it very carefully! The answer is 1.

An easier way to multiply larger numbers if you don't have a calculator.

Extras

Explain to your student why this method works using the distributive property.

a. $105 \times 62 = 105(60 + 2) = 630 + 210 = 840$

b. $58 \times 15 = 58(10 + 5) = 580 + 290 = 870$

c. $26 \times 31 = (25 + 1) \times (30 + 1) = 25(30) + 25(1) + 1(30) + 1(1) = 750 + 25 + 30 + 1 = 775 + 25 + 5 + 1 = 806$

d. $(3/8)/(4/7) = (3/8)(7/4) / (4/7)(7/4) = (3/8)(7/4) = 21/32$

Can your student visualize or estimate what % 16 is of 32?

Some number types had names and are still named today.

Prime numbers: 2,3,5,7,11 .... These types are only divisible by 1 and itself.

Ask your student to list the primes less than 100. Why are there less than 50?

Perfect numbers: 6 is the first one. These numbers have the property that the factors add to the number. 6 = 1+2+3.

Ask your student to find the next perfect number. (It is less than 30.)
          (Answer is 28.)

Comment: Some parents may be interested in the book by J. Davis, *Biblical Numerology*.

Challenge Problem

The ten digits can be written using tooth picks. Many c phones and computer use this method

    Example: The digit four can be written using 4 tooth picks

Can your student create the other 9 digits using tooth picks?

Chapter 4. The High School Prep Years - Grades 6-9
Section 4a. Algebra Formulas and Equations

A new technology does not add or subtract something, it changes everything.

Neil Postman

In a way, you might say using algebra takes the guesswork out of the problem! The word "algebra" comes from the Arabic "al-jabr" meaning "restoration," and refers to the fact that what you do to one side of an equation you must do to the other side to "restore" the equality. This brings us back to the words you were asked to define, formula and equation.

Definition: An equation is a statement indicating two numbers are equal.
> Example: a. $4 = 16/4$ or $4 = 4$.
> b. $3x - 7 = 2x + 3$, (implies) $x = 10$

Definition: A formula is a general rule stated as an equation representing a statement such as a theorem.
> Example: The formula for the area of a circle is $A = \pi r^2$.

Since the word area was used, then what is Volume? Your student probably thinks he knows, but ask your student to explain volume. The Concept of volume is an area moving through a height.

The rules for solving equations

1. If given an equation, then you can add a number to each side of the equation.

Comment: This permits subtraction since subtraction is adding the opposite, and also the postulate includes algebraic expressions since they represent numbers.

2. If given an equation, then you can multiply each side of the equation by the same non-zero number.

Comment: Dividing is the inverse of multiplication and dividing by zero is "out" since $1/0$ is not a logical. Can you justify that $1/0$ is not logical?

Assume you can divide by 0, then $1/0$ equals some number.

Example: $1/0 = N$ where N is the answer.
    Multiply by 0 implies $1 = 0N$ or $1 = 0$???

Ironically, the solving of an equation is not the difficult part. Arriving at or writing the equation is the most difficult and is what most students find the hardest. This process of writing the equations is the most important part in solving problems. Naturally, we will start with the easiest type of equations first to build your student's confidence.

List some formulas that your student has used or studied to date.

Here is a start:
    $A = LW$ $P = 4s$ $P = 2L + 2W$ $C = 2\Pi R$

Again, as a reminder, the early mathematicians were very capable and their students had the same problems that students have today. Here is an example.

What time of the day is it, if the time that is left today is 4/3 of the time that has passed?

Hint: Let x equal the time now, then the equation is:
    Time left (24-x) = (4/3)x. (time passed)
This equation reads the time left equals (4/3) the time of day.
        Answer: (10:17 A.M.)

Now do the assignment.

Chapter 4. High School Prep Years - Grades 6-9
Section 4b. Direct Variation

MANY OF THE LAWS OF THE SCIENCE ARE STATED IN THE LANGUAGE OF VARIATION

Unknown

You no doubt have noticed the relationship between various factors, such as the type of coat you wear depends on the weather, or the amount of water you drink depends on the type of activity? There are many such relationships. Here are a few more. Ask your student for a few more.
    Heating bill to outside temperature
    Cooling bill to outside temperature
    Take home pay to hours worked
    Take home pay to pay per hour
    The perimeter of a square to the length of its side

The miles per gallon to the speed you drive
Shoe size to height
Waist size to a person's weight
Skill to the practice time
Grades to the amount of time you study
Ask your student for a few more.

You can no doubt list many more. The point is, there are many variables that are related or connected in our everyday situations. The problem is to connect them quantitatively, in other words mathematically, if possible. You no doubt are thinking; How can this be done? First, you have to determine if the relationship is a direct one or not. This section will study only one type, which is called direct variation. One way to intuitively tell is to ask yourself if the two variables react in the same manner or direction. Take the first one listed above with regard to the heating bill and outside temperature. If the outside temperature increases, does the heating bill increase? In the second case: If the outside temperature increases, does the cooling bill increase? Which of these do you think represents direct variation? The answer is the second one since the two variables react in the same manner or in this case, both increase.

Comment: Discuss the other listed cases with your student.
(Classify each as direct or indirect.)

In order to do the mathematics for direct variation we must have a definition.

Direct Variation Definition: If two variables are so related that for each value of y there is an x such that y = kx, then x and y vary directly. k is the constant that relates the two variables. (This also indicates y/x = k, so you can always solve for k, the constant.).

Note for the parent: It is advisable to read this material before you help your student.

Example: Take an auto travel map where 1 inch represents 3.5 miles. Do you think this is direct variation between the distance on the map and the actual distance? In other words, the farther apart on the map indicates the farther the two points actually are a part.

Answer: Yes, it is direct, as one variable increases the other variable increases. If you answered yes, then when applying the definition we can write M = iK (i for inches) where M is the actual distance and i is the distance on the map.

The scale on the map reads: 25 miles = 1 inch. What is the k value?   M=iK

	Miles	inches	k(constant)	
	25	1	?	(.286)

(Do see where this value for k came from?)

		Miles	inches	k	
a.		7	2	?	(3.5)
b.		35	10	?	(3.5)
c.		?	21	3.5	(73.5miles)

A high school friend of the author's had a mid-1930s, Chrysler Air Flow, below.

Now back to the problem.

Hence: The formula is M = 3.5I. Since we know the formula, we can solve for either variable if we are given the value for the other variable. What happens to M, if the value for I is doubled? Tripled? quadrupled? This is the value of direct variation, being able to predict what will happen when one variable changes by using the formula.

Notice the procedure: First attempt to determine if the variables are related directly. Then if they are, write the formula (y = kx) using the definition and attempt to determine the value of k. The next step is to apply the formula to the problem.

Now work on the assignment from the teacher.

Two more examples for the student:
1. Do you think the circumference of a circle varies directly as the diameter? If yes, then write the formula using the definition.

71

From the following data, below, calculate k.

C	D	k
23.55	7.5	?
143.18	45.59	?

Answer: k = 3.14 (Rounded) "Eureka" you solved it!

2. Which of the following do you think vary directly?
  a. The distance a car travels and its speed.
  b. The braking distance of a car and its speed.
  c. The stress on the people in the car and the speed.
  d. Smoking and your health.
  e. Your grade on a test and the amount of time you study.
  f. Drinking alcoholic beverages and your ability to drive.
  g. Happiness and the amount you earn.
  h. The pitch of a guitar string and its tension.
  i. The area of a square and its length of side.
  j. The altitude of an equilateral triangle and the measure of its side.
  k. The wind pressure R on an automobile and the speed.

Answers: Take time to discuss the above.
   Vary Directly: a, b, c, d, e, h, i, j, k.
   Unknown: g

Chapter 4. High School Prep Years - Grades 6-8
Section 4c. Graphs: Line graphs, y = Mx + B

THINKERS RECOGNIZE WHEN TWO VARIABLES

ARE RELATED, BUT IT IS MATHEMATICS THAT CONNECT THEM NUMERICALLY.

Unknown

Note: Use centimeter graph paper, ruler, pencil or pen. Some of this may be a review for the parents also. The objectives are listed.
> 1. Be able to determine the x and y intercepts for a line.
> 2. Be able to graph linear equations.
> 3. Be able to graph linear inequalities.
> 4. Be able to graph linear equations involving absolute value symbols.

In one of their earlier geometry classes, your student may have learned that two points determine a line and through a point not on the line there is only one line parallel to the given line. These two conditions will come up in your student's high school math classes.

In this section, the student will graph equations that turned out to be straight lines and also will be given Descartes' definition of a line ($y = mx + b$).

Note for parent: Review this material before you help the student. It may be new material for some parents.

Questions to ask your student.
> If given the equations for two lines, what are the possibilities for their graphs?
> > a. Could the two lines intersect in one point?
> > b. Could they intersect in two points?

c. Could they intersect in no points?
d. Could they intersect in many or an infinite number of points?

Answer: All are yes except b.

Point out the following **in your house** to illustrate the following:

a. 3 Lines (really line segments) that intersect in one point. (a corner of a room)
b. Identify a plane
c. Lines that are parallel and also identify a plane
d. Lines that are not parallel and do not intersect.
e. Three lines that appear parallel and not in the same plane. Can your student identify some other plane segments in the house?

Comment: The name for the case (d) where lines don't intersect and are not parallel are **skew lines!** Check a dictionary.

Thinking Activity

Ask your student if two lines perpendicular to a third line will ever intersect?

After thinking about it show the student a globe and point that two lines perpendicular to the equator do intersect at the North Pole.

You may wish to explain to your student that the shortest distance on a flight is the arc of a great circle and not the straight line.

Chapter 4. High School Prep Years - Grades 6-8
Section 4d. DESCARTES' GIFT

...the essence of plane analytic geometry (Descartes') lies in the matching of ordered pairs of real numbers with points on a plane.

Edna E. Kramer
THE NATURE AND GROWTH OF MODERN MATHEMATICS

Rene' Descartes (1596-1650) invented the relationship between the geometric line and the algebraic equation. It has been said, that this creative invention "constitutes the greatest single step ever made in the progress of the exact sciences." (See E.T. Bell's *MEN OF MATHEMATICS*, page 35)

Suggested Reading: For a very readable account and report of the Descartes' interesting life and his accomplishments, see chapter 3 in Bell's *MEN OF MATHEMATICS*. (This book is in all libraries.)

From geometry, you no doubt recall that two points determine a line. Try it on your paper. How many points make up a line? Again, from geometry, we agree the answer is uncountable or infinite.

What Descartes did was to define a point as a set of ordered pairs of numbers (x,y) in the following manner. Graphically as follows: Label the horizontal number line x and another number line perpendicular to the x line label it y. The two lines intersect at the point zero on each line. A point is located by the number pair (x,y) where the x number is always first, an easy way to remember is to think alphabetical order. On the figure below, the point (3,1) is located by first locating 3 on the x line and then going up 1 on the y-axis.

```
         Q 2      Y  Q 1

       (-3,1) .        . (3,1)
       <————————0————————> X
       (-3,-1) .        . (-3,1)
          Q3            Q4
```

Q stands for quadrant and the X and Y axes are the real number lines.

The system is clever since if you know the values for x and y in (x,y), then you can locate the point. A similar system is used in your community to locate houses, businesses, etc., except along with the number a street name is usually given. The number pair system would actually be simpler but not quite as dignified. The above is not new to you since you worked with the system in your algebra and geometry classes. The Q1 to Q4 refers to the agreed upon system for numbering or identifying the quadrants.

Comment: One effective method of learning is by a continued review and then delving deeper into the subject after each review.

Notice the difference, in your student's geometry course, the terms **line** and **point** were classified as undefined. Where in Descartes' geometry a point and line are defined (below).

Definition: A point is a set of ordered pair of numbers written as (x,y).

Definition: A line is defined as the set of points that satisfy the equation y = mx + b, where x, y, m and b are real numbers with the restriction that x and y both cannot be zero.

If an equation can be operated on, using the postulates, or changed to fit this form, then it fits the condition in the definition and the graph is a straight line. Also, if you are given two points, then the equation for the line can be derived.

The activities for this section will refresh your memory and skills in graphing and deriving equations for straight lines. Sometimes the variables are other letters than x and y, especially in formulas used in the sciences. Two examples are C = $2\pi r$ for the circumference of a circle and A = lw is the area of a rectangle.

Activity for Understanding

1. Locate the following points on the x-y plane.
  a. (-4,-2) b.(-1,5) c. (-4,2) d. (-3,2) e. (-3,5)
  f.(-4,2) g. (2,2) h. (-1,-2) i.(1,5) j. (1,-2)
  k.(2,5) l. (3,-2) m. (3,5)

2. Graph the lines for the following three equations on the same set of x-y axes.
  a. y = 2x + 1 b. y = -2x + 1 c. y = 2

Select two values for x and two for y and what is the value for m and b in the equation of your line, $y = mx + b$?

3. How is the equation for the line determined from any two given points?
   Step 1. Write the general equation for the line. $y = Mx + b$. Substitute the given values for X and Y. (1,2) (2,5)

   Step 2. $Y = Mx + b$ $2 = 1M + b$

   Step 3. $Y = Mx + b$ $5 = 2M + b$

   Step 4. $5 = 2M + b$ and substituting $2 - m = b$ for b in step 2.
   $5 = 2M + (2-M)$ or $5 = 1M + 2$ or $M = 3$ and if $M = 3$ then $b = -1$ from step 1, $y = -x + 3 -1$ x -1.
   a. What are the x and y intercepts for each line?
   b. Pick any two points and what is the ratio of the y value of 1 point to the x value of the second point? Answer is 1.

4. What is the easy way to graph a line? How many points are needed? (2)

Comment: Discuss the intercept method as an easy way. Two points are needed, but you may want to use three points as a check.

5. How can you tell if a point, (x,y), is on the line?

a. Is (4,5) on the line for y = 3x – 7? Yes
b. Is (5,4) on the line for y = 7x – 3? No
c. List four points that are on the line for y = -2x + 6.
d. Are these two lines parallel? Y = 3x – 1, y=3x-5. From these 2 equations what do you think makes them parallel? (The value of m, is the same in each case the 3.)

Now help your student do the assigned homework.

Descartes was a Frenchman, where did he die? What is his famous statement?

"I think, therefore I am."

A Thinking Game: The Tower of Hanoi

A good game for the parent-student team to try to solve. The original problem consisted of the following set of 3 rings, a small, medium size and a large and you can add additional larger rings to make the game more difficult. The object is to transfer the rings from the starting post to another post so that the rings are in the same order. A move consists of only one ring at a time.

The rule: A larger ring can never be placed on top of a smaller ring. Coins of different sizes could also be used!

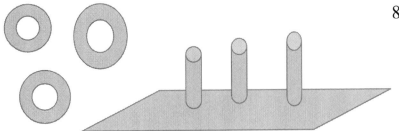

Start with 2 rings and the number of moves is 3. Then start with 3 rings and the number of moves are 6? Look for the related pattern and the formula for M (moves) given the number of rings. Instead of rings, coins can be used.

$M = KR$
$3 = 2R$ (3 moves when you start with 2 rings)
$? = 3R$
$? = 4R$
$? = KR$ Is their a pattern?

Chapter 4. High School Prep Years - Grades 6 - 8
Section 4e. The Addition Method
(To solve for the point where two lines intersect.)

RELATIONSHIPS BETWEEN DIFFERENT SUBJECTS (OR BRANCHES OF MATH) ARE CREATIVELY IMPORTANT IN MATHEMATICS.
Simon Singh

In the student mathematics courses to date, they have been solving very simple equations and possibly a few inequations in one or two variables. This understanding and skill is needed and applicable to any profession or vocation they wish to enter. Below are a few examples of formulas used in the occupations indicated.

$A = lw$ — Area landscaping or decorating
$V = lwh$ — Volume used in air conditioning and heating
$D = rt$ — Distance rate time relationship for forms of transportation
$S = 4\pi r^2$ — Surface area of a ball - sports
$A = p(1+r)^n$ — Compound interest formula used in the business world.
$S = .5gt^2$ — Distance an object falls in t seconds – aeronautics
$T = 2\pi(L/g)^{.5}$ — Pendulum clock formula - engineering
$W = VA$ — Formula for DC electrical power – engineering
$W = VA\cos\Theta$ — Formula for AC electrical power – engineering
$1/F = 1/d + 1/i$ — Optometry
$F = V_1 f/(V_1 - V_2)$ — Doppler sound principle – weather forecasting
$N = 10\log(I_2/I_1)$ — Formula for loudness in decibels-rock concerts – acoustics

This is probably enough to convey the idea that equations are used in everyday life, but we just don't realize it. This section is devoted to the various ways to solve systems of equations. You will be able to select the method that you feel is the most efficient or the easiest for the problem. In order to select a preferred method, you need to understand all four methods.

The addition method as you would guess employs the operation of addition and the use of the postulates, which basically says whatever operation you perform on one side of an equation you also perform on the other side.

Why do we need these ways to solve systems of equations? The answer is to have easier ways to find precise answers instead of approximate answers which you may have been reading off the graphs.

Comment: "Walk" with your student through the first problem.

Solve the following by using the above postulates as indicated.

1. Given two equations of lines: $2x + 2y = 5$ and $2x - 5y = 10$ multiply the second equation by -1 and rewriting the two.
$$2x + 2y = 5$$
$$-2x + 5y = -10$$
now adding the two equations results in $7y = -5$ which is an equation the student can solve and $y = -5/7$

The system has now been converted to a simple equation in one variable, which can be easily solved for y. In other words, the key is to convert the problem to one you know how to solve.

How is the value of x calculated? Just substitute the value of y into either of the original equations and solve for x.
$2x + 2(-5/7) = 5$ ---> $x = 6.3$ (you may need to help the student.)

a. The point is (6.3, -.7) Why is the point answer in decimal form?
b. What quadrant is the point in? (Q4)

Comment: The decimal answer in many cases is not exact since it is rounded off. An easy estimated check would be to graph the 2 lines.

If you notice, the procedure is to alter the problem to a type you feel comfortable solving. This is a procedure used throughout mathematics. Before we look at the other methods, let's take a break and consider a new kind of number.

Break time Question

Ask your student if the shortest distance between two points on the earth is a straight line? (Answer: It is an arc of a great circle.) Do you have a globe?

Chapter 4. High School Prep Years - Grades 6-8
Section 4f. A New Kind of Number

Number rules the universe.
The Pythagoreans, ca. 500 B.C.

A new type of number may create some interest or curiosity in your student and perhaps the parent and teacher. Recall in one of the history bits it was mentioned that Pythagoras had a Math Club and it evidently was a secret one. The story is that one of the members revealed a secret and suddenly he disappeared. Well, like most secrets they eventually become common information. The secret was that the square root of 2 cannot be written as the ratio of two integers, an example is 5/10 can be written as ½ (which is the Ratio of 2 Integers). The original thought was that all numbers were what is called rational, A/B (integer/integer). The $\sqrt{2}$ is not equal to A/B. The possibilities are, it either is or is not. The assumption was that it is rational, naturally, that was the only numbers they knew. So, $\sqrt{2}$ is equal to E/O (even/odd) or O/E (odd/even) or O/O (odd/odd). With that given can you show that $\sqrt{2}$ is not equal to any of these and therefore the $\sqrt{2}$ is not rational.

Hint: Assume the case $\sqrt{2}$ = Odd/Even and now square each side.

What do you have and is it a contradiction? $2 = (O/E)^2$ or $E^2(2) = O^2$ Which says an even number is equal to an Odd number!

Can you reason the other two cases with your student?

(This method is the indirect method. The indirect method is to list all the possible conclusions and justify all are impossible but one, hence the one that is left is the answer.)

There are other types of numbers also, like imaginary.

A unique problem for the parent and student, which will impress others with your student's ability to solve for the sum of a set of integers.

Question: What is the sum of the first 10, or 20, or 30 odd integers? Or, what is the sum of 1+3+5+7+9+ ...+30?

The key is in the way you approach the problem!

Cases	The problem	The sum
1	1	1
2	1 + 3	4
3	1+3 + 5	9
4	1+ 3 + 5 + 7	16
5	1+ 3 + 5 + 7 +9	? What do you think the sum is?
N	1+3+... +N	?

The objective in this type of activity is to observe the items and try to see a pattern between the cases and the sums.

The sum is the square of the items in the "cases", which means the sum of the first 10 is 100, 20 is 400, and 30 is 900. The sum of the first n odd integers is N^2.

Another **Indirect Proof** to re-enforce meaning and method.

Galileo is known for showing that objects heavier than air fall at the same rate.

Here is his reasoning: The possibilities are that Heavier objects fall faster than Lighter objects, or Lighter objects fall faster than Heavier objects, or they both fall at the same rate. (These are the only possibilities.)

Assume the Heavier object falls faster, then tie the Lighter one to the Heavier one and the new mass should fall faster since it is heavier, but the Lighter should slow the heavier one down and therefore the combined heavier object should fall slower.

A CONTRADICTION

You have the same type of contradiction if you assume the lighter object falls faster.

Hence: The lighter and heavier fall at the same rate, the only remaining possibility!

He then went to the leaning tower of Pisa and showed the public his conclusion.

Research: Galileo, Life and works (very interesting)

Chapter 4. High School Prep Years - Grades 6-8
Section 4g. The Substitution Method

MATHEMATICS IS LIKE A MIGHTY OAK WITH COUNTING NUMBERS FOR THE ROOTS. ARITHMETIC GROWS ON NUMBERS, ALGEBRA ON ARITHMETIC, AND GEOMETRY ON ALL THREE.

Unknown

Another method used to solve systems of equations is the substitution method. It is very similar to substitutions in athletics. Except in math the subs are equal. An example will illustrate this for you.

If given the following system, and you are to find the values of x and y which will satisfy both equations, or in other words where the two lines intersect.

Line 1 is $3x - 4y = -6$ and line 2 is $2x + 5y = 19$.

Using the substitution method equation (you may pick either equation), selecting the second equation which can be transformed to:

$2x = 19 - 5y$ and therefore: $x = 19/2 - 5y/2$ by dividing by 2.

Now we know what x is and substituting this value for x into the first equation we have:
$$3(19/2 - 5y/2) - 4y = 6$$
This is an equation you can solve since it has only one variable.

Which of the following values is y equal to? (1, 2, or 3)
Answer: 3

Once you know the value of y, then substitute the value for y into either equation and solve for x. Which of the following values is x equal to? (1, 2, or 3)
Answer: 2

Naturally, there are other possible methods, such as you could have solved for y first and then for x. This method permits you to use your own ingenuity to solve the equations.

Diophantus (300 B.C.E.) wrote one of the oldest treatises on algebra. When do you think he lived? (See comment at the end of this section.)

Understanding the method

Solve for x and y by the substitution method.

1. $x - 2y = 11$
   $2x - 3y = 18$

2. $3x - 4y - 12 = 0$
   $5x + 2y + 6 = 0$

Answers: 1. (3,-4) Quadrant IV   2. (0,-3) (Point on the y-axis)

3. Check your answers to 1 and 2. by graphing the lines.

Now help your student with teacher assigned problems!

Special type of problem!

A Diophantus Equation

Mathematics is not a careful march down a well-cleared highway but a journey into a strange wilderness, where explorers (the students) often get lost.
W. S. Anglin

This is an example of the type of problem known as a Diophantine Equation or problem. They are very interesting and also classified as thinking problems.

Old problem

A farmer purchased 100 fowl for $100. Roosters cost $5, hens $1 and chicks 10 for a dollar. A few questions to help you understand the problem.

   a. What is the maximum number of roosters? (Less then 20-why?)
   b. What is the maximum number of hens.
   c. Why will the number of chicks always be a multiple of ten?
   d. How many of each did he buy?

   Hint the equations are:
   $R + H + C = 100$
   $5R + 1H + 19 C = \$100$

Many attempts to solve this problem using Algebra, but give up when they realize they have more variables than equations. After thinking about the situation they realize the additional unknown or condition is that the answers are integers, so now the problem can be solved. (Reduce the two equations to one equation in 2 variables, and use the condition that the 2 variables are whole numbers.)

   Answers: a. $R < 20$. b. $H < 100$. c. Always a whole number. d. 9, 51, 40 (Answer)

Comment: Diophantus (330-236 B.C.E.) did not work with negative numbers. Dates are now listed as BCE defined as

"Before the Common Era" and CE for Common Era. The year is 2017 C.E. Some people object to this method.

Do a computer search for more information on Diophantus.

Diophantus died at the age of x, but from a translation of his epitaph we know the following:
- 1/6 of his years as a child
- 1/12 of his years as a teenager
- 1/7 of his years as a bachelor
- Five years after his marriage his son was born.
- The son died four years before Diophantus did.
- The son lived half the years Diophantus did.

Write the equation: X/6 +X/12 +....
   a. How many years did Diophantus live?
   b. At what age did he marry?
   c. At what age did the son die?
       Answers: 84, 42, 38

Chapter 4. High School Prep Years - Grades 6-9
Section 4h. Graphing Short Cuts

MATHEMATICS IS NOT A SPECTATOR SPORT
                                   NCTM CONVENTION

Team effort for the parent and student. The objectives to understand:
  1. Be able to determine the x and y intercepts for a line.
  2. Be able to graph linear equations.

In one of your student's courses, the student learned that two points determine a Euclidean line and through a point not on the line there is only one line parallel to the given line. These two conditions, may have been classified as postulates. What is a postulate?

Research: What is the Euclid's famous parallel fifth postulate? Do you recall? In your geometry class a line was classified as an undefined term?

Euclid's 5th Postulate: Through a point not on a line, only 1 line can be drawn that is parallel to the given line.

In this chapter, you graphed equations that turned out to be straight lines and also were given Descartes' definition of a line ($y = mx + b$). Your student may need to review their notes. In this section, you will investigate the graphs of what we call systems of equations, inequations or linear inequalities, and how they are related to geometric figures.

If given the equations for two lines, what are the possibilities for their graphs?
    Ask your student.
    a. Could they intersect in one point?
    b. Could they intersect in two points?
    c. Could they intersect in no points.
    d. Could they intersect in many or an infinite number of points?

    Answer: All yes except b.

Point out lines (really line segments) in the classroom (or a room in your home) to illustrate the following:
    a. Lines that intersect in one point.
    b. Lines that are parallel and identify a plane.

c. Lines that are not parallel and do not intersect.
d. Three lines that are not parallel and are not in the same plane.
e. Three lines that intersect, but are not in the same plane.
f. Can a line be perpendicular to 2 lines and all three intersect in a point.
Hint: The corner of your ceiling.
Comment: The name for the case (d) where lines don't intersect and are not parallel. (skew lines! Check a dictionary).

1. Graph the lines for the following three equations on the same set of x-y axes.
   a. y = 2x + 1   b. y = -2x + 1   c. y = 2x - 1

   (Easy points to graph are when x = 0 or y = 0. These are called the x and y intercepts.

   The graph for 1a:

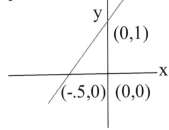

   The slope for "a" is 2. (In this case, a positive slope.) (M = 2)
   a. What do you observe about value of M and the direction or slope of the lines?
                +∕ or -∖ related to M
   b. What are the x and y intercepts for each equation?
   c. What are the slopes for each? (a is +, b is -, c is +) _

Answers: Two of the lines are parallel. (a, c)
(You may need to review the definition of slope.)
Intercepts: (0,1) (-.5,0), (0, 1), (.5,0), (0,-1), (.5,0)
Slopes: 2, -2, 2
2. What is the easy way to graph a line? How many points are needed?

Comment: Discuss the intercept method as an easy way. Two points are needed, but you may want to use three points as a check. Select the intercepts as the two points. (Let x=0 and then y=0.)

In the equation Y = MX + B, what does M tell you if:
    a. M is +? b. M is –?   c. there is no M value?
    d. if M is zero?
In the equation Y = MX + B, what does it tell if :
    a. B is 0?   b. B is +?   c. B is –?
Now work on the teacher assigned problems.

Question. Can you find a point in your house where 3 lines intersect in a point and are not in the same plane? Hint: Look up at a corner.

Another Parent-Student Thinking Problem
Camp Problem

The cook at the summer camp needed 4 quarts of water to make orange juice, but he had only 3, 5, and 8 quart containers. The cook asked you to take the containers and bring back 4 qts. You figured it out and recorded how to do it for the future. Write down the procedure!

Solution

Moves	8	5	3		Moves	8	5	3
1	8	0	0		1	8	0	0
2	5	0	3		2	5	0	3
3	?	?	?		3	5	3	0
4	?	?	?		4	3	2	3
5	?	?	?		5	6	2	0
6	?	?	?		6	6	0	2
7	?	?	?		7	1	5	2
8	?	?	?		8	1	4	3
9	?	?	?		9	4	4	0

If your student has a different order, pat the student on the back and call (him or her) Descartes!

A Decision Making Problem

Bank A charges $3 service charge per month plus $.10 for each check you write.
Bank B charges $4 service per month plus $.05 for each check you write.

    a. Explain under what conditions you should use bank A.
    b. Explain under what conditions you should use bank B.
    c. Explain under what conditions it doesn't matter which bank you use.
    d. Graph the cost line for each bank to show the answers to the above questions.

Answer: The charges are equal at 20 checks, therefore the bank selection depends on how many checks you think you will write.

Comments and review

Chapter 5. Introduction to Statistics - Grades 9-10
Section 5a. Averages

STATISTICAL THINKING WILL ONE DAY BE AS NECESSARY FOR EFFICIENT CITIZENSHIP AS THE ABILITY TO READ AND WRITE.

H. G. Wells

The Meanings of the Three Averages
(Mean, Mode and Median)

Thoughts to motivate your students.

The calculating of averages is fairly simple, but interpreting the results accurately can be confusing and is often misleading. You may come across numbers representing averages in the news every day and as an informed citizen you should be able to understand the correct meaning of these. You have probably heard the statement: "Figures don't lie, but liars figure." It is important for you and your student to understand the use and misuse of these averages and their implications in order for valid interpretation. This chapter section will help interpret statements involving statistics and other information that is needed to see the valid or invalid picture.

The reporting of scores or sets of data to convey the results or the "picture" by using the term average is often used.

Some examples of these are:
    a. The meaning of math scores for a certain grade or class in your community.

b. The average weight of the Green Bay Packers defensive line.
c. The average score per game for your school's basketball team.
d. The average stopping distance for a certain make of car.
e. The average speed for a trip.
f. The average for your school's ACT or SAT scores.
g. The age for the average voter with regard to an issue.

Let's say you are driving from Missoula, MT to Billings, MT a distance of approximately 350 miles and it is all interstate highway. On this section of freeway there are no stoplights so once on the freeway you can set the cruise control and not stop until you are at your destination, unless you need a rest stop. Given the following graph, explain why the average speed was approximately 50 mph, and only slightly higher if the 20-minute rest stop is not included in the time. You may need to use the following definition.

Definition: The average speed is the total distance divided by the time it takes to travel the distance. $S = D/T$

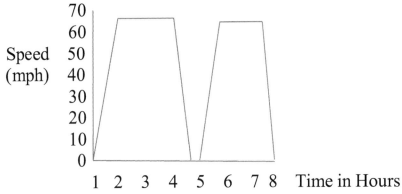

For your student: From the graph, estimate the time the car was actually traveling 65 miles per hour? Where is the rest stop indicated? What could the driver do so that the average speed is nearer the speed the cruise control is set at? What is your estimated average speed for the trip? What would the average speed have been without the rest stop (44mph)? If you left at 8 A.M., then what was the time for the rest stop?

This car is a mid-1930s Chrysler Airflow. The author had a high school buddy whose father owned one.

Let's look at the arithmetic average of the following set of scores on an eighth grade 2014 math team exam.

Scores: 100, 99, 80, 76, 70

The superintendent proudly reported to the newspaper that the average score for the eighth grade 2014 math team is 85. Does this one number give a true picture of the team's scores? Notice, no student even scored an 85.

Definition: The MEAN or arithmetic average for a set of numbers is the sum of the numbers divided by n, the number of items in the set.

$$\text{Formula: Mean} = (\text{sum of n scores})/n$$

Calculate the mean for the above eighth grade math team set of scores and check if 85 is correct.

The 2015 and 2014 math team scores are listed below. The superintendent reported the team had the same average in each year.

    2014 scores: 100, 99, 80, 76, 70
    2015 scores: 100, 95, 91, 80, 59

Calculate the mean score for each year.

For both years it was reported an 85 average was reported to the parents. Is that correct? Do you think the two teams are the same with regard to their accomplishment? Which team would you say is better overall? Should the parents be satisfied with just one number reported for each of the two teams?

Comment: Discuss the above situation with your student. Now to further confuse the interpretation of data, there are three types of averages: **mean, mode** and **median**. Each is used by the business world, in education, the news media, the financial world and in the medical world. The mean was defined above and the mode and median will now be defined.

Definition: The MODE for a set of data is the most popular or most frequently occurring element or score in the set.

Definition: The MEDIAN for a set of data is the middle element or score when the elements or scores are arranged in order of size or magnitude.

These three, (mean, mode and median) are each called the average, but the one that is being used is not always reported. Each of these averages, will be further clarified and compared, by the applications.

Here is a set of income figures representing the annual payroll in a small business, including the owner's salary.

$18,000, $18,000, $18,000, $22,000, $25,000, $25,000, $80,000

Which income does your student think is the salary of the owner? Why?

Answer: Probably $80,000

Using the three definitions calculate each average. The
Mean is ____.
The Median is ____.
The Mode is ____.

Answers: Mean = $29,428.57. Median = $22,000.

Mode = $18,000

Each of these could be reported as the average!
Which average do you think the owner claimed for the average pay? Why?

> Answer: Mean or $29,428.57

Which average do you think the employees claimed for the average pay? Why?

> Answer: Possibly the mode, $18,000.

What is the mode and median for each of the 2015 and 2016 math team scores.
  2015 scores: 100, 99, 80, 76, 70
  2016 scores: 100, 95, 91, 80, 59

> Answer: No mode in either set, but the Median is 91 for the year 2016 team and 80 for year 2015.

Team Activity for parents and student

In the following calculate and identify the three averages. Use your calculator and/or computer. Team work usually provides better understanding! (Your calculator may have these programs built in, or even your computer.)

1. First arrange the scores in ascending order, and then calculate the three averages.
   a. 3, 5, 7   b. 0, 1, 3, 5, 16   c. 0, 1, 1, -2, -3
   d. 2, -3, 6, -7, -1, -4, 7
   e. 71, 65, 98, 57, 64, 69, 87, 88, 79, 48, 77, 80, 50, 75, 85, 90, 99, 30, 94, 96, 80.

Answers:
1a. Mean = 5, mode (none), median = 5
1b. Mean = 5, mode (none), median = 3
1c. Mean = -3/5, mode = 1, median = 1
1d. Mean = 0, mode (none), median = -1
1e. M = 1582/21 = 75.3, Mode = 80, Median = 79

2. Many times, scores are grouped for easier calculating of the averages and for graphing. (Graphing of scores will be studied in the next section.) In the following case, the scores are grouped in sets of 5.*

Score	Number of students
96-100	5
91-95	9
86-90	10
81-85	13
76-80	14
71-75	15
66-70	13
61-65	10
56-60	7
61-65	3

   (*The Actual scores, 96,97,98,99,100, are not listed but there are 5 in the first set as indicated.)

   a. How many students took the test?
   b. What is the sum of all the scores? Assume the scores are at the center of the interval. (5(98) + 9(93) + 10(88) + ... + 3(63).
   c. What is the mode score?
   d. What is the median score?

e. What is the mean score?
   Answers: a. 100   b. 7650   c. 73   d. 78   e. 76.5

*The actual scores which were assigned to the intervals in problem 2 are listed below.

With the computer or even with some calculators the true measure for the averages can be easily calculated using built in programs. There is no need for the schools to report only the mean score for a class or school. Actually, the whole graph should be reported and the three averages indicated. This is the correct way to really report the true picture of test scores like the ACT or SAT and then use the tests to **improve** the curriculum and teaching areas.

Student Scores*
100,98,99,97,97,96
95,95,94,94,94,93,93,92,91
86,86,87,87,87,88,88,88,88,89
85,85,84,84,84,84,84,83,83,83,82,82,81
80.80.80.79,79,79,78,78,78,77,77,76,76,76
71,71,71,72,72,72,73,73,73,73,74,74,75,75,75
66,67,67,67,68,68,68,68,68,68,68,69,69
61,62,62,62,63,64,64,64,65,65
56,56,57,57,57,58,60
51,53,55

Questions
   a. How many students took the test?
   b. What is the sum of all the scores?
   c. What is the mode score?
   d. What is the median score?
   e. What is the mean? (Use your calculator.)

Answers: a. 100 b. 7642 c. 68 d. 76 e. 76.4

3. Montana is a very large state, approximately 700 miles long and 400 miles wide, yet the evening national weather news may assign one number to indicate the temperature for the whole state. The following numbers indicate the variation of temperatures comparing five of the larger cities in Montana in August. What are your students comments when the news media reported these temperatures for the State of Montana.
   101, 95, 67, 45, 36?

   Questions:
   a. What is the mean temperature for these readings?
   b. What is the median temperature for these readings?
   c. What is the mode temperature?
   d. What would you report for the state of Montana?

   Comment about the answers: There is no correct figure for the State's temperature.
   a. Mean is 68.8   b. Median is 67.
   c. There is no mode.
   d. Just report the high and low and where these are.

4. Montana Baseball players keep track of their batting averages. A player has an average of 290 after 75 times at bat. What will he have to hit in the next 40 times at bat to have an average of 300?
   Answer: 318.75
   Equation is $300 = [290(75) + 40x]/(75+40)$
   What is the definition of x?

5. A baseball player's batting average is 280 in 30 times at bat. How many hits will he need in the next 80 time at bat to average 300 for the 110 times at bat? Estimate the answer before you calculate the correct answer.
   Answer: Approximately 30 (H=30.2, but it is impossible to hit a .2 hit)

6. Which is the better class taught by the same teacher? Class A has a mean of 85 and a range of 40, Class B had an average of 85 but a range of 20. Which do you think would be easier to teach? Why?
   Answer: Class B is the better class. In theory, the smaller the range the easier the class is to teach.

Thinking Activity

How many counting numbers less than 20 can you write using 4 fours?
   Examples:
   $1 = 44/44$ $2 = 4(4)/(4+4)$,
   $3 = (4+4+4)/4$ $4 = 4\sqrt{4} - \sqrt{4} - \sqrt{4}$
   $5 = \sqrt{4} + \sqrt{4} + 4/4$ $6 = 4+4-(4/\sqrt{4})$
   $7 = 4+4 - 4/4$ $8 = 4+4/(4/4)$
   $9 = 4+4 +4/4$ $10 = 4(4) - 4 - \sqrt{4}$
   $11 = 44/(\sqrt{4}+\sqrt{4})$ and so on!

Your notes and comments

Chapter 5. Introduction to Statistics - Grades 9-10
Section 5b. A bit of the History of Statistics

You cannot fake in mathematics, no one can be fooled. You can either prove (or solve) or you cannot.

Jerry P. King
THE ART OF MATHEMATICS

Even Aristotle was interested in Statistics and I would guess the concepts of the terms mean, median and mode were used back in the "early days". Statistics was really placed on the math investigation list in the 1800s and became applied during world War 1 as you will see in the following example. The military needed shoes for the men and naturally could not wait to measure each man's feet and then place the order. The following graph will illustrate how the problem was solved. Take the time to explain to your student why the curve is called the normal curve and how it helped the military determine the number of shoes and other material in order to equip the men when they were drafted. The Mean, Mode, and Median are called Measures of Central Tendency. Babbage in the 1800s invented an interesting calculating device which to a degree was the forerunner to the computer. (Suggestion: a computer search for Babbage.)

Somewhere I read a cartoon which went like this:
> Student A. "My father has a better bowling average than your Dad. In the Monday nights league his average is 210. In the Tuesday nights league his average is 90. Thursday night its 205, and on Friday the average is 193.

Student B.   My father stays home nights and we work on Math.

The Normal Curve

The curve is called normal because so many items in nature fit the curve. An example in nature is the curve for sunset time for a year. The height of the boys or girls in your school. The larger the school the closer the curve is to normal. The curve for alternating electrical current is somewhat normal. Educational testing scores for large groups if displayed or graphed will be near normal. Plant growth and products can be normal in various ways like, height, size of product, weight of product. The trend is stated by Neil Postman in the following quote.

"Just as statistics has spawned a huge testing industry, it has done the same for the polling of "public opinion."

Activity for Parent and Student

Take 5 coins and toss all five as a set and count the number of Heads in **each toss**. Record the number of heads for each of the 20 tosses. Do this each day for a week and record the totals for the 7 days. Then construct a graph to show the results. You may be surprising.

Totals are the number of times for: 5 Heads, 4 heads, 3 heads, 2 heads, 1 head and 0 heads resulting from the 140 tosses.

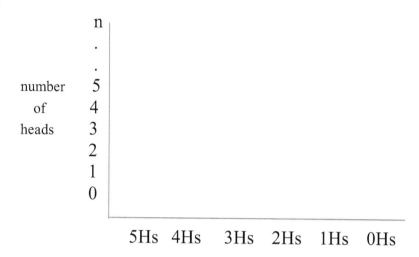

The curve may be similar to what is called the Normal curve below.

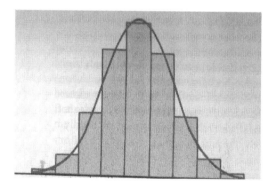

Interesting Project

Perhaps the school administration would provide the school's SAT scores for last year and your student could construct the graph and observe if they fit the normal curve. There will be more on this topic in Chapter 8.

Chapter 6. High School - Grades 9-10
Section 6a. Why Study Mathematics?

The study of mathematics is an investment in the future!
Unknown

Comment: This chapter will provide a better understanding of your student's background, and better prepare the student's future background. As Music Man's Prof. Hill would say, "You gotta know the territory."

Read the chapter and select the activities you feel your student needs for future courses or a review of past courses.

The first two sections are short of a review of concepts your student previously studied and some skills requiring not only recall but also explanations as to why. There are a few unusual thinking type problems and other investigations. The reason for this comment is to caution you against a possible first impression that this chapter is to elementary or to difficult.

Is your student able to make decisions and what does he base them on? Is your student able to justify his decisions, and logically present and defend his decisions or positions on issues? The author believes all students can think and make valid logical decisions, but they need to be taught how. Mathematics is a bridge to a better future, but it must be taught for the how and why of decision making with everyday non-math examples.

You are probably thinking this is an odd way to begin the study of a mathematics section. The author's aim is to help you, the parent and teacher, to create some curiosity.

The attitude of the students has about a course will, to a large degree, determine their success. Attitude does make a difference! You must remember a grade is nothing more than a measure or indicator with regard to past performance. It may have nothing to do with your future accomplishments. (Ray Kroc dropped out of high school. Who was Ray Kroc? **Comment:** Ray Kroc, founder of the McDonald food chain, grew up in Oak Park, Illinois.)

In all your studies or classes, you should keep asking "why". It was Einstein that said: *"The important thing is to never not stop questioning..."*

The former chairman of the World Trade Corporation advises youth who are hopeful for a bright future to study mathematics for its bi-products.

YOUNG PEOPLE WHO HAVE ACQUIRED THE INFORMATION, PUT THE PIECES TOGETHER TO FORM TENTATIVE SOLUTIONS WILL ALWAYS BE IN DEMAND.

J. G. Maisonrouge
Board Chairman
IBM World Trade Corp.

In a former chapter you were able to rapidly determine the sum of the first n odd integers. What is the sum of the first 20 odd numbers? Remember? (400)

Now can you determine the sum of the first 20 even integers?

Recall how to look for a pattern?

Cases	numbers	sum
1	2	2
2	2+4	6
3	2+4+6	12
4	2+4+6+8	?

What is the pattern so you can predict the sum for?
    20   2+4+6+8   ?
    N   2+4+6+8 ... N ?
Answer is 420 for 20 and $N(N+1)$ for N.

Chapter 6. Why Study Mathematics? - Grades 9-10
Section 6b. What is Mathematics?

GOD gave us the integers (whole numbers) and all the rest is the work of man. (and Women.)
L. Kronecker

It has been stated that mathematics began when man began putting together bits and pieces of math to meet the needs in various applications such as building, measuring, banking and navigation.

Students often ask: Why do I need this math course? This is a valid question. The answer is, all too often, just study now and some day you will understand and realize why. But most of the time it takes an unusual and truly motivated person to study for the pure quest of knowledge. If students can be convinced from the beginning that this course will be practical, useful, applicable, and enjoyable, then the positive attitude will be easier to achieve. Mathematics is a bridge for the following areas, which were predicted to have the brightest future in the first few decades of the twenty-first century. The brightest future means the ones, which will have the greatest opportunities for your selected employment. All require skills involving various areas of mathematics and in different amounts. (Order is not relevant.)

    Health Services
        Primary-care workers
        Nutrition counselors
        Gerontology specialists

    Hotel Management and Recreation
        Restaurant workers
        Resort employees
        Travel agents
        Conference planners

    Food Services
        Managers and chefs
        Food processing
        Food laboratories

    Engineering
        Robotics
        Aviation
        Construction

    Waste management

Computers
- Maintenance
- Design
- Retails sales

Consulting services

Business Services
- Accounting
- Statistical analysis
- Payroll
- Word processing

Human Services
- Personnel relations
- Job evaluation
- Benefit planning
- Legal assistance
- Public relations

Financial Services
- Insurance
- Financial planning
- Banking

Teaching
- Day-care
- Primary
- Secondary
- Science
- Mathematics
- Computers

Foreign language

Maintenance and Repair
    Industrial
    Home
    Business

The study of mathematics will give you some of the tools needed for the above along with your other courses involving the skills of:

Writing   Drawing   Communicating   Key boarding
Computer literacy   Decision-making   Science
Team participation or getting along with others.

Chapter 6. Why Study Mathematics? - Grades 9-10
Section 6c. Why is Geometry different?

We learn the new in the light of the old.

                                            Unknown

Finally, the ancient mathematicians began to put it together or organize the conclusions into a logical system. The Greeks really started this organization during what is called their Golden Years. They also were the first to logically build a geometry course in their academies. Socrates and Plato started schools for adults (male) who may become the leaders of the country. At the entrance to Plato's Academy was the sign:

LET NO MAN ENTER HERE IGNORANT OF GEOMETRY.   *Plato 400 B.C.E.*

(Presidents Clinton and G.W. Bush established a modern day version called the Presidential Leadership Scholars Program in 2014 or15 to develop future leaders.) They have had 3 graduation classes (80 students per class) already.

Your student may ask WHY? (Geometry teaches a person to become a valid thinker, a need for a democracy!)

A logical system is built and the basis is the following: **Undefined terms, defined terms, postulates, followed by conclusions** (called theorems in math and laws in your state). **Jefferson's Declaration of Independence** is a great example. Read with your student the first few pages to observe the above. You will recognize the undefined terms, the defined terms, the postulates or assumptions and the conclusion that the colonies should be free. **(This is why Plato and Euclid wrote the first Geometry text and why colleges and high schools today require the study of geometry. The problem is most high school Geometry courses do not emphasize the relationship with applications to decision making in non-geometric world.)**

Look at the last 2 sentences in the above paragraph, the ones inside the ( ). Do any of the words need defining? I'll assume the answer is yes and ask which words need defining? You can't understand a sentence completely if it contains words which need defining. List the words that your student classifies as undefinable?

Classify each word in the sentences in ( ) suggested in the above paragraph as defined or undefined. You may be surprised! (complete the following)

How many are: Undefined —— Defined ——
   Total # of words?   ——
   What % are Undefined?   Defined ?
      (U = ?% (T#W)

Comment: Take time to discuss the words that need defining with your student. Pick a few sentences from any book or better take the Pledge of Allegiance and classify each word as defined or undefined.

Approximately 30-50% of the words in any sentence are undefinable.

Recall: A definition is valid if when it is reversed it is still true.

Examples:
1. A dog is a four legged animal. Is this a valid definition? No! Why?
2. A square is a 4 sided figure. Is it a Valid definition?

Discuss a few postulates or assumptions from everyday experiences. Some such postulates or assumptions are:
   The sun will rise in the east.
   The electrical power will be on.
   The car will start.
   The water for the shower will be warm.

List with your student additional assumptions and a few decisions based them.

If you know the definition of postulate and said it was an assumption, then you are correct. All games consist of a set of rules, and are summarized in a rule book to insure fairness and play ability. The rule book consists of the postulates, definitions, and explanations of how these terms are used to determine the play.

Example: Let's take the Saturday game of football since I'll assume the time of year is Fall. (Why would that probably be a valid assumption?) There are a large number of needed definitions for clarification, the players, the play of the game, the time of play, the scoring, and the penalties. You can probably name many more. Check a game rule book.

We make assumptions in life every day. The selection for the clothes we put on each day are based on the assumption: We think we know what the weather will be.

Definition: A postulate is a statement, which is assumed to be true.
Definition: The set of real numbers is the union of the rational and the irrational numbers.

Comment: You may have to define union, but be sure to understand the set of reals are all the numbers on the number line.

To understand this definition, we need to define rationals and irrationals.

Definition: A rational number is a number, which can be written as the ratio of two integers.
What are a few of these different types of rational numbers?

The Real Number System, is harder to understand. One way to think of it is to draw a line like the one below and for each point on the line assign a number.

```
<--------------------/---/---/---/---/---/---/----------------->
         <--- -3 -2 -1  O  1  2  3 --->
```

Now imagine all the types of numbers you can put on the line in order of size. Ask your student to list some types.

Comment: It may help to actually draw the line or do it on paper and fill in a few integers, fractions, decimals, radicals, and hopefully you will recognize the set of real numbers is the union of rationals and the irrationals.

Definition: An irrational number is a real number, which cannot be written as the ratio of two integers. (Recall the case of the square root of 2.)

List a few other irrational numbers?

Postulate: There is a one to one correspondence or matching between the points on a line and the real numbers.

Comment: This is an important postulate, which permits us to apply algebra to geometry and the reverse.

Postulates for the Operations with Real Numbers.

Comment: You may recall these 11, but it helps to review and explain them for further understanding. (Number refers to real numbers) Make a copy of these for future use.

1. Add or subtract any two numbers and you always get an answer.
2. Multiply or divide (except division by 0) any two numbers and you will get an answer.

   Mathematicians label the above 2 as the **Closure Postulates**.

3. For any two numbers A + B equals B + A. (Addition is commutative.)
4. For any two numbers A x B equals B x A. (Multiplication is commutative)

   Mathematicians label 3 & 4 the **Commutative Postulates**. What does commutative mean? Are subtraction and division commutative?

5. For any two numbers A + (-A) = 0
6. For any two numbers A(1/A) = 1, providing A is not equal to zero.

   Mathematicians label 5 & 6 the **Inverse Postulates.**

7. For any two numbers A + 0 = A.

8. For any two numbers A x 1 = A.

> Mathematicians label 7 & 8 the **Identity Postulates.**

9. For any three numbers A + (B + C) = (A + B) + C.
10. For any three numbers (A x B) x C = A x (B x C).

> Mathematicians label 9 & 10 the **Associative Distributive Postulates.**

11. A(B+ C) = AB + AC

> Mathematicians label #11 the **Distributive Postulate.** (Notice it changes a multiplication problem to an addition problem.)

Ask your student these questions:

1. Is putting on your socks and shoes on a commutative operation?
2. Is starting the car and the putting the key in to start it a commutative operation?

These 11 are the postulates for real numbers and definitions make up the basics for the beginnings of a logical system and permit the use of algebra to solve geometry problems.

Geometry Postulates For a
Logical Geometry Decision Making Course

(These postulates are the ones for the author's geometry book. The set may be a bit different for another Textbook.)
1. A line has an infinite set of points.
2. Two points will determine one and only one straight line.
3. There is a one to one correspondence between the points on a line and the real number system.
4. The shortest distance between two points is a straight line.
5. The length of line segment is the is measure of the shortest distance between the two points.
6. The shortest distance between a point and a line on a plane is the measure of the perpendicular line segment (not valid in all geometries.)
7. Three non-collinear points will determine a plane.
8. If two parallel lines are crossed by another line (transversal) then the alternate interior angles are equal in measure.
9. Through a point not on the given line, there is only one line through the point that is parallel to the given line (in plane geometry).
10. The area of a rectangle is equal to the length times the width and the answer is in sq. units.
11. The formula for the area of a circle is $A = \pi r^2$
12. The formula for the circumference of a circle is $C = 2\pi r$.
13. The volume of a rectangular solid is the length times the width times the height. $V = lwh$

Ask you student these questions:
1. Is brushing your teeth and putting on the paste on the brush a commutative operation?

2. Is reading a book and opening it a commutative operation?
3. Is putting your coat and cap on a commutative operation?

Comment: There will be more on the how as to the importance of everyday Decision Making methods and conclusions.

These are the building blocks for a logical system which we call Geometry (Plane and Solid). It was Plato, Socrates and others who recognized that this was the foundation for critical thinking or logical DECISION MAKING and why Socrates started his academy for logical reasoning for future leaders. (A democracy depends on it.) This is why schools teach Geometry.

Add your Notes

Interesting Problem

1. What is the following message?
   ICUR ICUB ICUR2YS4ME
   Hint: I see you are . . . . .
2. The figure below is not drawn to scale. From the indicated size of the angles, which segment is the shortest? Be able to justify your answer. Hint: In a triangle the largest side is opposite the largest angle.

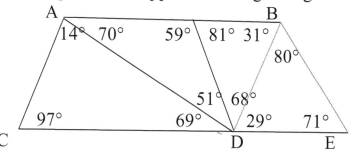

Answer: Side DE

Area problem

3. How can this be? (Each figure, below, is composed of the same figures.) The figure on the right is made from the figure on the left.
   What is the area of each figure? How can this be?

   Perhaps copying one of the figures, cut it apart as indicated and assemble the other figure will help you see the error.

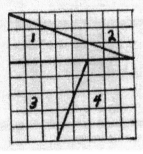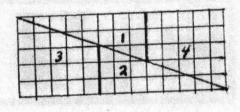

Chapter 7 International System of Measurement
Section 7a. The World's Measuring Units

The Advancement and Perfecting of Mathematics are closely joined to the Prosperity of a Nation.

Napoleon

(Check with the teacher for any assignments on this section.)

A Short History of the Metric System

After the concept of the Metric or SI (System International) was establish it took over a hundred years for it to be adopted by most of the world countries. The United States still has not switched over in all businesses. Industries and companies doing business in the international market have switched, but not in all the local applications. Many areas use both in the labeling of the products. Therefore, your student needs to be able to understand how to switch from one to the other.

One of the arguments was the use of the decimal system for simplification. The noted scientist and mathematician, Fibonacci (1000 A.D.), urged the use of the decimal (base 10) system, but Simon Stevin (1600) is given the credit. The problem was how to write the decimals. The need for a standardized system worldwide really became needed from about 1700 on. The foot may have been the length of the kings foot, a yard could be the length from the nose to the tip of the longest finger. At one time, it was suggested to have a 10day week, a three week month, but a 12 month year. Finally, with the help of scientist-mathematicians and the leaders of mostly European countries, the meter was define as a 1/10,000,000 of the distance from the equator to the North Pole. (Gauss in the early 1800s urged the adoption of the SI system of measurement, but it took another 100 years to be adopted internationally. all countries. Why Gauss? He was the most respected mathematician in the world at that time. Suggest a computer search as to SI.))

Linear conversions
(Use the equality relationship.)

1. How many meters is the 100 yard football field?
   From the conversion table, we know:
   1 yard = .9144 meters or multiplying by 100 we have 100 yards = 91.44 meters.
   Another way to work the conversion is 1yd/91.44 M = 100 yds/ x meters, and solve for X.

2. Some states list their speed limits in both MPH and in Kilometers (KPH). What is 60MPH equal to in KPH?
   From the table: 1MPH/1.609KPH = 60MPH/ n(KPH) and again solve for n. The answer for n is 96.54 and rounding the answer to 100 it is printed on the speed signs as 60mph = 100 KPH.

3. An area problem: Your house may be rated as 1600 sq. ft. What is the area in sq. meters? 1 sq. ft./.092903 = 1600sq ft/n sq. m., or n equals 148.64 sq. m.

4. A ranch in Montana has 1000 acres, the buyer from Europe wants the answer in hectares.
   1 A/ .405 H = 1000/ nH
   The above reads as: One acre is equal to .405 Hectares as 1000A /nH and solve for n. What is your answer?

Select a few other useful conversions for your student to work with like weight or height. (The number of sq. meters on the bedroom's floor.)

Chapter 7. International System of Measurement
Section 7b. SI Conversion Table

English		Metric or SI
1 inch (in.)	=	2.54 centimeters (cm.)
.39 in.	=	1 cm. =.1 decimeter (dm.)
12 in. = 1foot (ft.)	=	30.48 cm.
3 ft. = 1yard (yd.)	=	.9144 m.
39.37 in.	=	1 meter (m) or 100 (cm.)
5280 ft. = 1 mile = 1 league	=	1.609 kilometers (km)
.621 mile	=	1 km.

Area

1 sq. ft. = 144 sq. in. = 929.03 sq. cm. = .092903 sq. meters
1 sq. yard = 9 sq. ft.   =   .836 sq. meters
1.196 sq. yd. = 10.765 sq. ft. = 1 sq. meter
1 sq. miles = 640 acres   =   259.004 hectares
2.471 acres   =   1 hectare
1 acre = 43560 sq. ft.   =   .405 hectare

Volume (Area through a height)

1 cubic in.   =   16.388 cubic cm
1 cu.ft.=1728 cu.in.
1 cu.yd.   = 27 cu.ft.   =   .765 cu.meters
1.308 cu.yds = 35.31 cu.ft. = 1 cu. meter
1 U.S. fluid Gal=4qts = 8pts = 128fluid Oz.

1 U.S. gal. = 3.785 liters
1 British gal. = 4 liters

Question: (Why does a gallon of gas in Canada cost more than a U.S. Gallon? It is larger.)

A few conversions for the kitchen
1 teaspoon = 5 milliliters (ml)
1 tablespoon = 15 ml
1 oz. = 29.6 ml

Thinking Problem for Parents and Student
The 10 Coin problem

Arrange 10 coins and number each one as below.
    1 2 3 4 5 6 7 8 9 10

The object is to end up with 5 piles of 2 coins each. Sounds easy, but the rules are:

1. You can select a coin and you can **only jump 2 coins** in either direction and land on the third one.
   Example: coin 1 can jump two coins and lands on 4.
   After the first move from above the coins are:
     2 3   4 5 6 7 8 9 10
            1

2. Once a pile has 2 coins it is out of play, but a pile of two can be jumped.
   In the case above, 5 could jump to 3 and the result is shown below.
       2 3 4   6 7 8 9 10
           5   1
   Now the play is "dead" since there is no way to jump using coin 2. (See rule 1.)

So start over! Remember 5 plies of 2 coins each is the objective. It can be done! What usually happens a person will make the proper sequence of moves, but will forget the order because it wasn't written down. Have fun!

3. The following data is the result of interviewing a group of parents. 13 were driving Ford Products, 21 were driving General Motors Products, 15 were driving Chrysler Products, 4 were driving only Fords and GMs, 2 were driving all three products, 5 were driving only Fords and Chryslers, 3 were driving only cars by General Motors and Chryslers.

Draw the Venn diagram (see below) depicting the above information. (Use a similar diagram to the one below and fill in the regions from the given data with the numbers.) Hint: Ask your student where the area is that represents owners that own all three makes.

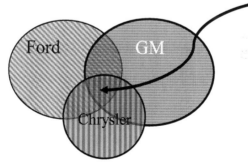

Questions: How many were driving only Fords? Only GMs? Only Chryslers?

Answers: Only Fords = 2, Only GMs = 12
Only Chryslers = 5

Chapter 8. High School Years – Grades 9-12
Section 8a. Plane and Solid Geometry Review

(This chapter assumes your student has had geometry.)

Let no student exit (high school) here ignorant of Geometry.

J. Elander

Understanding evolves from work, appreciation is from applications.

Unknown

This Geometry content will be reviewed as a separate review since the major tests like the SAT or ACT results have indicate Geometry has one of the lowest scores and the least understood. It is assumed your student has had a year of basic Plane and some Solid Geometry or is taking Geometry.

Short answer questions-use a ruler for drawings

Parent and Teacher Suggestion: Let the students justify their answers in class.

129

The answers are given to this review. Keep track of the missed problems, explain the how and repeat the missed questions in a few weeks.

Remind your student: Don't forget to use the 5 indexes for quick reference to assist you in the solutions. Be able to justify or explain your answers, and you may even have to explain your answers to your parents or request their help! The related theorems are listed in ( ) if help is needed.

All valid conclusions, theorems in mathematics, are based on Undefined terms, Defined terms, Assumptions, and previous Conclusions.

Geometry Review
(Use the Indexes)

1. Are the following statements valid definitions? Defend your answers. (A definition is valid if the definition is true when reversed.)
    a. A restaurant is a place that serves food.
    b. Mathematics is a useful course.
    c. A postulate is an assumption.
    d. A plane triangle is a set of three non-collinear points and the line segments determined by the three points.
2. What are the three geometric terms in 1d that are classified as undefined?
3. How many points are needed to determine a geometric line?
4. The points on a line correspond to the numbers on the _____ _____ line.

5. Draw a ray and label it AB.
6. Draw a line segment and label it AC.
7. Draw a triangle and label it CDE.
8. A geometric plane is determined by ___ ___ ___ points.
9. What is the sum of the angles in a plane triangle? (Theorem 3)
10. How is the distance between points A and B determined?
11. What is a theorem?
12. What are the conditions for 2 triangles to be similar? (Theorem 7)
    a.   b.   c.
13. When are triangles congruent?
14. a. What are parallel lines?
    b. What are skew lines? (Need a dictionary?)
15. Draw three acute scalene triangles, label each ABC. (Use your ruler and protractor. Theorems 21,22,23)
    a. In one triangle draw the three medians.
    b. In the second triangle draw the three altitudes.
    c. In the third triangle draw the three angle bisectors.
    d. Write a conclusion for each.
16. The shortest distance from a point to a line is the ___ distance.
17. If A implies B and A is given, then ___.
18. Draw a rhombus that is not a square. (Need a dictionary?)
19. Draw: a. convex polygon. b. concave polygon (Need a dictionary?)
20. If the sides of a triangle are 46, 23 and x, then what do you know about the measure of side x? (Theorem 1)
21. The number of square units a plane figure contains is its ___.

131

22. The Pythagorean Right Triangle Theorem states _____. (Theorem 18)

23. What is the sum of the interior angles in each of the following figures? (Do not use a protractor. Hint: In a triangle the sum is 180 degrees.) Do you see a relationship between the number of sides in the figure and the angle sum?)

a.   b.   c.

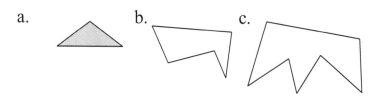

d. What is the sum of the angles in a 12 sided polygon?

24. If each letter represents a unique digit, then what are the possible digits for the following addition problem?

```
    G O
  + G O
  -------
  W I N
```

Hints: a. Which digit is W?
b. G must > than ___
c. The O can't be ___

25. How many planes may four points determine?

26. If A implies B, then does Not A imply Not B? Must explain your answer!

27. What is the perimeter of right triangle ABC, where C is the right angle? Given: AB = 3x units and BC = 4x units

28. What is wrong with this argument?
    A yard is 36 inches.
    ¼ yard is 9 inches.
    (Take the square root.)
    Therefore, ½ yard is 3 inches.

29. Students of school X voted that all students will wear a RED cap at the Saturday game. Which of the following are valid? Hint: If-Then form.
    a. John wore a red cap, therefore he is a student of X.
    b. Joe is not a student of school X, therefore he will not wear a red cap to the game.
    c. Jim did not wear a red cap to the School X game, therefore he is not a student at school X.
30. Draw a sphere inscribe within the cube is a sphere. The cube has a side equal to s. (Theorems 30,31,32,34)
    a. What is the volume of the cube?
    b. What is the volume of the sphere?
    c. What is the surface area of the cube?
    d. What is the surface area of the sphere? (Theorems 35,36 and 37)

Answers to Review

1. a. True, but not a definition. b. True, but not a definition. c. Yes, a definition. d. Valid
2. Point, line, and plane
3. Two points
4. Real numbers
5. A————→B
6. A————B
7.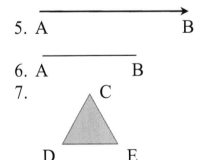
8. Three non-collinear points

133
9. 180 degrees
10. Absolute value of A-B
11. A theorem is an important mathematical statement that can be proved.
12. a. Angles are equal (AA).
    b. The corresponding sides are proportional (SSS).
    c. Sides proportional and included angles equal (SAS).
13. Triangles are congruent if they are similar and the ratio of sides is 1.
14. a. Parallel lines are lines in the same plane and do not intersect.
    b. Skew lines are lines not in the same plane and do not intersect.
15. Each intersect in a point
16. Perpendicular distance
17. Then B
18.   19. a. ▭ b.
20. 23 < x < 69 units
21. Area
22. In a right triangle with sides a, b, and c, then $a^2 + b^2 = c^2$ where c is the hypotenuse.
23. a. 180   b. 540   c. 900   S = (n-2)180   d. 180 degrees
24. 4, 0
25. Four: ABC, ABD, ACD, BCD
26. Converse is not true
27. 12x units
28. Listen to their explanations
29. "c" is valid, the contrapositive
30. a. Vol. of cube is S^3 cu. Units
    b. Vol. of the sphere is $(1/6)(\pi S^3)$ cu. Units

c. Surface area of cube is 6s²sq. Units
d. Surface area of the sphere is πs² sq. units

Calculate your percentage correct. (C% = [#Correct/(36)] times 100.

Suggestion: If your students missed any of these problems, record the missed problem numbers, then help the student correct them and understand their errors. Record the problem numbers hat your student missed, wait a week or two and retake the review. This chapter has 11 sections and when you complete them your final test score will amaze you.

(The proof of the pudding is in the eating.)
What does the above statement mean?

There is nothing wrong with missing a problem once or twice, but....!

Challenge Problem

Given in the figure below:
    a. Angle A is 40 degrees
    b. AC = AB, OC = OB
    c. OC bisects angle C

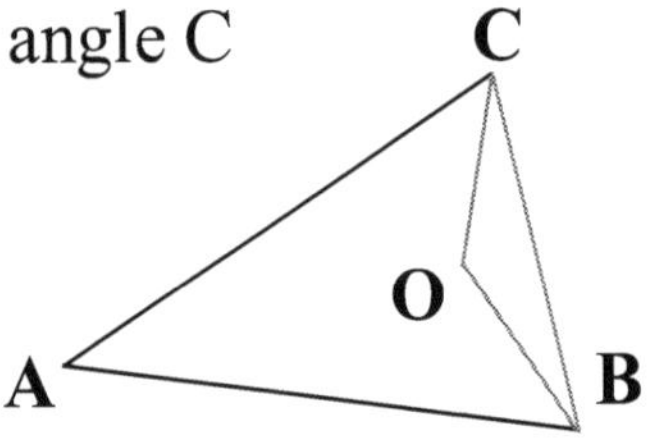

What is the measure of angle BOC?

Answer: Angle BOC is 110 degrees

Chapter 8. The High School Years – Grades 9-12
Section 8b. Algebra Summary

Methods for Solving Quadratic Equations and understanding inequalities

Mathematics: The Queen of the Sciences

E. T. Bell

(Parents and teachers should read this chapter carefully, since it possibly contains material you have forgotten.) **Your student will need a scientific calculator, a ruler, protractor and graph paper.**

In chapter 4, your student solved and worked with equations and their graphs that were straight lines, plus inequalities where the graphs were regions or areas. The equations had either one or two variables, and were classified as first degree. By first degree equations mathematicians mean that the exponents for the variables are 1. (Since the exponent is 1, we do not write the 1. It is understood to be 1.) In other words, if the exponent is not indicated, we define it to be 1. It has been mentioned that the "game" of mathematics is different from most games. It gets more difficult as it becomes more advanced. The equations your student has been solving were all first degree. The next type will be second degree equations involving one or two variables and at least one of the terms will have an exponent of 2. Two examples are $y = Ax^2 + Bx + C$ or $k = xy$. The second example is the case students usually question as a second degree. It is a logical question. Notice the phrase at least one term will have an exponent of 2. The term xy indicates multiplication and when the operation is multiplication the exponents are added, therefore the **term** xy has a power of two and that makes it a second degree. It is even more convincing when you graph K = xy. It is a curve, not a straight line. This case will be explained later. Previously the types of equations were basically y = Mx + B and the graph is a straight line, where M is the slope and B is the y-intercept.

In this section the type of equation $y = Ax^2 + Bx + C$, will be explained for the A, B and C related to the graph, plus the methods for solving this type will be explained. (You know, after some thought, what point C is on the graph!)

Comment: C is the y-intercept when x equals zero.

Definition: A Quadratic equation with two variables is defined as $y = Ax^2 + Bx + C$, where A, B and C are rational numbers and A is not equal to 0.

Some equations are classified as functions, which your student probably defined in their first course in algebra, but the definition will be repeated here.

Definition: An equation is a function if the set of ordered pairs (x,y) satisfies the condition that for each possible x value there is only one y value.

Notation: $Y = Ax^2 + Bx + C$ is a function in x and can be written as $f(x) = Ax^2 + Bx + C$. The letter in the ( ) indicates the letter that is used as the variable.

$f(T) = 5T + 39$ is another example. This could be read as the function T is equal to $5T + 39$. The value of the function at 2, written $f(2)$, is $5(2) + 39$ or $f(2)$ equals 49. The case for the quadratic equation, we would write it as $f(x) = Ax^2 + Bx + C$ which is a function.

The question now is: how will the student solve quadratic equations? (Is there an easier way?) **Remember, this is just a few "hints" so the parent will know what the student is learning and what the teacher is presenting.**

The student will probably start with the various cases and progress to the solution for all cases! (This is probably a review for the parent of material you had what now may seems as years ago!)

How to solve the quadratic equations

(Each type will be explained for you, but be sure to understand the solution. So, you will have an understanding of what the student is learning.)

Type 1. $Ax^2 + Bx = 0$ (2 examples):

a. $x^2 + x = 0$
   $x(x + 1) = 0$ (Factoring)
   Why is the following correct?
   $x = 0*$
   or

b. $2x^2 + 4x = 0$
   $2x(x+2) = 0$
   $2x = 0$
   $x = 0*$
   or

x + 1 = 0   x + 2 = 0
x = -1   x = ?
Answers: 0, -1   x = 0, -2

*Reason: If the product of two numbers is zero, then at least one of the numbers must be zero.

Type 2. $Ax^2 - C = 0$ explained.

Comment: The case where $ax^2 + c = 0$ leads to what are called ($ax^2 = -c$) imaginary numbers (example $\sqrt{-4}$) How would you find the square of a negative number? Think about it). We will not introduce imaginary numbers, but the next math course your student takes will no doubt teach imaginary numbers.

The following two cases are completed, but be sure to understand the solution!

a. $10x^2 - 250 = 0$ b. $5x^2 - 20 = 0$
   $10x^2 = +250$ $5x^2 = 20$
   $x^2 = 25$ $x^2 = ?$
   $x = 5$ or -5 $x = ?$ or $?$

The method is to convert the equation of one of type 2:
$Ax^2 - C = 0$ to and $x = +$ or $- \sqrt{C/A}$

Note: The value for C/A must be positive at this stage of the game, since the student cannot take the square root of a negative number.

Comment: Recall that the square root symbol $\sqrt{}$ means the positive square root.

Type 3. Reduce the case to one of the above and factor.

Comment: Notice all second degree equations have two answers. The two answers may be identical. If types 1 and 2 don't fit, then try factoring the equation so that the two factors equal zero.

a. $x^2 + 2x + 1 = 0$
   $(x + 1)(x + 1) = 0$
   $(x + 1)^2 = 0$
   $x = -1$

b. $x^2 - 2x + 1 = 0$
   $(x - ?)(x - ?) = 0$
   ?

Notice: This really is a type 1 case, but involving factoring. What is the other answer for "a" and "b"?

Type 4. Completing the Square

If cases 1-3 don't fit, then the following method will work, but as you would expect, it is more difficult.
$$x^2 + 8x + 1 = 0.$$

One way to attack the problem is to try to convert it to one of the above types by adding a number to each side.
Example: $x^2 + 8x + 1 + 15 = 15$

The equation can now be written as $(x + 4)^2 = 15$ and taking the square root of each side results: $x + 4 = +\sqrt{15}$ or $-\sqrt{15}$ where $\sqrt{}$ means the positive square root.

Solving for x we have:

x = -4 + √15 or -.13 (rounded to 2 decimal places)
x = -4 - √15 or -7.87

Comment: There is an easier method to calculate the number to be added, but the general formula to solve all quadratics is easiest, so we will skip to type 5.

Type 5. The General Formula will solve all cases. (The formula is derived below!)

The general quadratic equation is $Ax^2 + Bx + C = 0$ and the two solutions are:

$$x = \frac{-B + \sqrt{B^2 - 4AC}}{2A} \qquad x = \frac{-B - \sqrt{B^2 - 4AC}}{2A}$$

The above formula will solve all quadratic equations. This is what you have been looking for and it makes the solution quite easy with a calculator. The variable does not have to be x, naturally! The following was used by Benjamin Peirce in the 1870s. His method was something like the following:

    Given: $Ax^2 + Bx + C = 0$
    Step 1: Multiple each side by 4a.
        Write your answer.
    Step 2: Subtract 4ac from each side.
        Write your answer.
    Step 3: Add B^2 to each side.
        Write your answer.
    Step 4. Factor the left side of the equation
        Answer is: $(2AX + B)^2 = B^2 - 4AC$.
    Step 5: Take the square root of each side.
        Answer: $2AX \pm B = \pm \sqrt{B^2 - 4AC}$

Step 6: Add -B to each side.
The result is?
Step 7. Divide each side by 2A

You should have this formula for the solution to $Ax^2 + Bx + C = 0$:

$$x = \frac{-B + \sqrt{B^2 - 4AC}}{2A} \quad \text{or} \quad x = \frac{-B - \sqrt{B^2 - 4AC}}{2A}$$

Problems for your student's understanding

Solve the following quadratic equations in one variable. When radicals are involved, give the answers in radical form (the exact answer) and then rounded to two decimal places for the approximate answers (use your calculator). The key is to alter the equation to fit one of the types explained above, or use the formula.

a. $m^2 - 6m = 16$ b. $8x^2 - 8x + 5 = 0$

Answers:

m = 8, -2 $x = 4 \pm \sqrt{11}$ or 7.31, -.68

c. If the answers to a quadratic equation are 5 and 7, then write the equation?
Answer: $x^2 - 12x + 35 = 0$ Hint: $(x-5)(x-7) = 0$

Quadratic equations are very useful as a means for determining the maximum or minimum values in following, TV signals, reflectors, and in the medical field, financial world and many more. The following example will illustrate minimum point.

On a sheet of graph paper graph $y = x^2$ for the integer values for x from -3 to +3.

Table: x = -3 -2 -1   0   1   2   3
       y =   9   4   1   0   1   4   9

Using graph paper plot these points and observe the curve. Do you see the minimum point?

It should look like this graph.

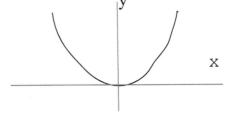

A parabola with a minimum point at the point (0,0).

Now graph $y = -x^2$ for the same set of values and you will have the graph for the maximum point. What caused the graph to flip or invert? It has been said that the parabola is one of the curves that controls our lives. The statement refers to the conic sections and their applications. There are 6 possible conic–plane sections, but only 4 are usually identified. The following is a list of items to support the meaning of that statement.

    Satellite dish   Automobile headlights
    Searchlights   Loud speakers
    TV &Telephone tower receivers and transmitters
    Path of an object thrown in the air, like a football
    Flashlight reflectors
    Some types of optical lenses
    Suspension bridges   Medical (operations)
    Radar   Reflecting telescopes

You are probably wondering why the second degree equations needs to be solved? The following problem is a simple explanation with an understandable, practical back yard problem.

Your family plans to have a garden in the back yard and purchases 100 feet of fencing. (You need fences in MT to keep the deer out.) The father says the rectangular shaped garden is to have a maximum square area with the 100 feet of fence. He recalls how to do it from his Algebra class in college.

The conditions:
  Perimeter is 100 ft. $= 2L + 2W$ or $50 = L + W$
  Area $= LW$ square ft. $A = LW$ or $L(50 - L)$

Where did the 50-L come from?

Now, graph $A = +50L - L^2$ or $A = -L^2 + 50L$
           Table of values:

Length	5	10	15	20	25	30	35	40
Area	225	400	525	600	625	600	525	400

From the table you can tell the maximum area is 625 when the length is 25. The parents may have known this was the maximum value from their college math and the max. value (or minimum) is the x-value at the turning point.

Turning point = [-B/2A, F(-B/2A)]

Notes
Add your comments and notes for understanding

Conic plane intersection means how many geometry figures may be formed when a plane intersects a double cone?

The curves resulting from a plane intersecting the double cone are: Point, Line, Circle, Parabola. Ellipse, and a Hyperbola. Can you see them? It has been said that these curves control our lives!

Chapter 8. High School Math - Grades 9-12
Section 8c. Graphing Inequalities and Applications

The connection between the improvement of the human conditions and happiness of the human race is Science.
Neil Postman

The easy way to graph inequalities is to first graph the equality, and then shade on the graph the inequality area by indicating one side of the line or the other side.

1. a. Graph the set of points that satisfy the inequality, $y > 1x - 1$
   b. First graph the line for $y = 1x - 1$.
   c. Shade or color the regions for the solutions to "a."

   Answer:

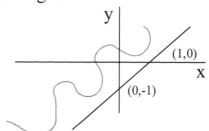

If this were a profit function it would be indicated by the set of points above the line as the profit area and below the line is the debt area.

Answer: For the inequality, the symbol < means the set of y < 3x + 1 values below the line y = 3x + 1. The symbol > means the set of y values above the line y = 2x + 1 or greater than.

What symbols tells you that the line is included?
Answer: ≥ or ≤

2. Graph on the same set of axes the set of points that satisfy the following inequalities.
   a. Y < -x + 3
   b. x ≥ 0
   c. y > x + 1.

Answer: a. Below the line. b. Above, plus the line. c. Above the line.

3. What is the solution to the following inequality? Hint: Graph on one set of axes.
   Y ≥ X² − 1 (shade the solution area.)

Now do the assigned homework!

As you may guess there is more than one way to solve systems of equations and inequalities. Mathematicians are always looking for easier ways to solve problems. The advances in technology have played a major role in this area. **Carefully read and understand the definitions, postulates and the theorems, since these are the keys to success.**

In the above, you practiced with your student solving systems of equations and inequalities by graphing. The graphing method is slow and the values for the solution point hard to read from the graph unless you have graphing calculator. The graph provides a picture or a common sense interpretation for the problem and a "ballpark" location for the answer.

This approach is to answer two methods: (1) graph the equations to get a general estimate for the answer. (2) use algebra to interpret the answer more precisely.

Like learning to play a new game, you have to read and understand the rules. It is the same in mathematics. Some people even classify mathematics as the most important game considering its role in and contributions to our quality of life. So as a review, we will start with some definitions and postulates. (rule book) The major objective for this section is for you to understand the "rules" and apply them.

Definition: A point is a set of ordered numbers written as (x,y).

Definition: A line is defined as the set of points that satisfy the equation: $y = mx + b$

Definition: Direct Variation. If two variables are so related that for each value of y there is an x such that $y = kx$ where k is the constant, then x and y vary directly. (K is equal to the ratio y/x.)

We also need a few new conditions. (Why are these called postulates?)

Postulate: If given a set of equations, then the equations can be added (subtracted) and the result is an equation.

Postulate: If given an inequality, then a number can be added (subtracted) to each side of the inequality and the result is an inequality of the same order.

Postulate: If given an inequality, then the inequality can be multiplied or divided by a positive number and the result is an inequality of the same order.

Notice as stated in the postulate above an inequality can be multiplied by a **positive** number. What happens if you use negative numbers?

Postulate: If given an inequality, then the inequality can be multiplied (divided) by a negative number and the result is an inequality of the opposite order.

Note: Does you student understand what "same or opposite order" means?

Postulate: If given a set of inequalities of the same order, then the inequalities can be added and the result is an inequality of the same order.

Like any sport, you need a lot of practice (Home Work) to really play the game well.

Activity for Understanding

In the following, ask your student to explain how he or she would work each problem.

1. On the same set of axes graph the following three inequalities and name the figure. (Answer: Area of triangle and ask the student to solve for the area?)
   a. $y \geq -1$ b. $y \leq 1x+2$ C. $y \leq -1x +2$

2. Graph these 2 inequalities and shade the solution area.
   $Y \geq X^2$ and $Y \leq 2$.

3. Thinking application: Which Bank? (Graph the two equations and then answer the questions.)

   Bank A charges $3 service charge per month plus $.10 for each check you write. Bank B charges $4 service per month plus $.06 for each check you write. The equations are:
   Bank A: $C = 3$ per month + $.10 n (n = the number of checks)
   Bank B: $C = \$4$ per month + $.06n (n = the number of checks)

       b. Graph the equations on the same set of axes and show the solution for **1 month** and **n** checks. Extend the lines until they intersect and beyond.
       c. Explain under what conditions you should use bank A.
          d. Explain under what conditions you should use bank B.

e. Explain under what conditions it doesn't matter which bank.

Discovery Problem for Parent, Teacher and Student

Your great grandfather left a note in his lock box which you finally found the key and opened it. In it was the following message and map: The gold is located at the spot that is equal distance from the pine tree, the huge rock, and the opening to the cave. How would you locate the point where to dig for the gold? See figure below.

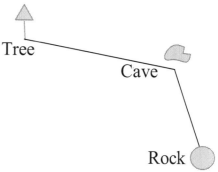

What geometry theorem will you use? Hint: Consider the 3 points as on a circle and you need to find the center of the circle. (Theorem 26)

If you are interested in another case, see chapter 31 or 34 in the very popular book *Treasure Island*.

Notes

Your student may find this interesting! Below is the formula to solve any quadratic equation and a way to check the answers.

$Ax^2 + Bx + C = 0$ and the 2 answers are from section 8b.

$$x_1 = \frac{-B + \sqrt{B^2-4AC}}{2A} \quad \text{or} \quad x_2 = \frac{-B - \sqrt{B^2-4AC}}{2A}$$

Ask your student to add the two answers ($x_1 + x_2$). The answer is -B/A. Answer is -B/A. Now help your student to multiply x_1 times x_2 and explain what the answer means. The answers to the equation $x^2 -2x - 8 = 0$ are 4 and -2. Now add the 2 answers. Ask your student what the value of -B/A is? (2/1). **This is an easier way to check if your answers are correct**.

Solve the equation and check your answers. **(Theorem 43)**
    a. $x^2 -4x -12 = 0$ b. $x^2 -6x +8 = 0$
    Answer: a. 6 and -2   b.= 4 and 2

Chapter 8. High School Math Years - Grades 9-12
Section 8d. Exponents

The miraculous powers of modern calculations are due to three (mathematical) inventions: Hindu Notation, Decimal Fractions, and Logarithms.
(What would you add from today's world? (computers)
                                          F. Cajorie
                             HISTORY OF ELEMENTARY
                                  MATHEMATICS(1896)

(A Easier Way to Solve Practical, but Difficult Problems)

You know what exponents are, but this section on exponents will prepare you for the next section on LOGARITHMS? This section will help your student understand this 400 + year old invention. Logarithms are really exponents and were a real time saver at that time. A by-product is the slide rule. Logarithms and the slide rule in the past played the same role as the calculator and computer does today.

Research: John Napier (logs) and Henry Briggs (slide rule).

The concept of logarithms is an important one for problem solving, and is needed more in today's math dependent world. (Perhaps the teacher, or a parent, will have a slide rule to show the class and explain how it is used.)

Exponents

This section will not only introduce the student to new material, but will follow more of a college type format, meaning you will have to be guided though the development. Some of the material may be a review, but most will be new.

The concepts and skills involved are used in the worlds of banking, physics, chemistry, geology, medicine, music, space travel and, of course, the world of mathematics.

The term exponent is one you are familiar with so we will start with it. (We learn the new in the light of the old!) From your Algebra back ground the following definition is one you will recognize.

Definition: a. If $x^m x^n$, where x is real and n and m are integers, then the product is x^{m+n}.

Example 1: 10^3 times $10^2 = 10^5$ or 10x10x10 times 10x10 = 10^5.

b. If x^m/x^n, where x is real and n and m are integers, then the quotient is x^{m-n}, where m > n.

Example 2: 10^3 divided by 10^2 = (10x10x10)/(10x10) = 10^{3-2} = 10

The definition from your Algebra class probably stated m and n were positive integers or possibly depending on your text they were required to be non-negative. We need a definition to satisfy the condition where m and n are real numbers.

The first case is where the exponent is 0. In other words, what is the value of a number like x^0, **when x is not zero**. The following will guide you to the answer.

Example 3: x^3/x^3 = xxx/xxx or 1, but from the above definition we know the answer is x^{3-3} or x^0. Since we know the first expansion is, it follows that x^0 must also be 1 (x cannot be zero).

The next case answers the problem when the exponent is negative such as 3^{-2}.

Example 4: $x^2/x^5 = 1/x^3$ this we understand. By definition we think the answer should be x^{2-5} or x^{-3}. We know that x^2/x^5 is the same as xx/xxxxx, which is the same as $1/x^3$; therefore, in order to be consistent mathematicians define 1/xxx to be the same as x^{-3}.

The third case will answer the problem when the exponent is a fraction or even a decimal.

Example 5: $(x^{1/2})(x^{1/2}) = x^1$ but we also know the $\sqrt{x}$ times the $\sqrt{x}$ is equal to x. Therefore: $x^{1/2} = \sqrt{x}$.

These conclusions will be stated as definitions as in most texts, but we could call them theorems, since we justified them.

Theorems: Properties of exponents
   a. If x^m times x^n, where x is real and n and m are rational, then the product is x^{m+n}.
      Examples: $10^5 \times 10^{-3} = 10^2 = 100$
                 $10^{1/2} \times 10^{-1/3} = 10^{1/6}$
   b. If x^m/x^n, where x is real and not equal to zero, and n and m are rational, then the quotient is x^{m-n}.
      Examples: $10^2/10^{-3} = 10^5 = 100000$
                 $10^{1/2}/10^{-1/3} = 10^{5/6}$
   c. x^0 is equal to 1 providing x is not zero.
      Examples: $10^{-3}/10^{-3} = 10^0 = 1$
                 $10^{-5}/10^{-5} = 10^0 = 1$
   d. If given x^{-n}, where x is real and not 0, and n is rational, then x^{-n} equals $1/x^n$.
      Examples: $5^{-2} = 1/5^2 = 1/25$

$$10^{-3} = 1/10^3 = 1/1000 = .001$$

Comment: This enables you to convert expressions with negative exponents to ones with positive exponents!

 e. If $x^{n/d}$ and x is real and n and d are rational and d is not zero, then the expression $x^{n/d}$ is equal to $\sqrt[d]{x^n}$ (which is read as the d root of x to the nth).

 The reasoning leading to this:
$\sqrt{2}$ times $\sqrt{2}$ = $2^{1/2}$ times $2^{1/2}$ = 2^1 hence, it follows $\sqrt{2} = 2^{1/2}.$

 Example: Calculate the value of $8^{2/3}$. First without your calculator, and then using your calculator.

 Answer: Cube root of 8 squared is 4 or it could be read as the cube root of 64.

 f. If $(x^m)^n$, where m and n are rational numbers, then the product is x^{mn}.

 Example: $(10^3)^2 = 10^6 = (10 \times 10 \times 10)^2 = (10 \times 10 \times 101)(10 \times 10 \times 10) = 10^6$.

Understanding Activity

Using the above definitions, simplify the following.
 1. 3^3 2. $(-3)^3$ 3. $-(-2)^4$ 4. $-(-1/2)^4$ 5. $(2)^3(2)^2$
 6. $5^7/5^4$ 7. $3^7/3^4$ 8. $5^4/(-5)^3$ 9. $(.5)^2(1/2)^4$

10. $3^5/(-3)^6$ 11. $(.1)^2/(.1)^4$ 12. 5^0 13. $4^{-1/2}$

Your notes and comments

Chapter 8. The High School Years–Grades 9-12
Section 8e. Logarithms

The invention of logarithms Laplace said amounted to "shortening the labours (and) doubled the life of the astronomer."

F. Cajorie
HISTORY OF ELEMENTARY MATHEMATICS

After completing this section, your student will understand:
1. Definition of logarithm.
2. Calculate a few simple problems using logarithms.

Logarithms, invented by John Napier (1550-1617), and modified by Henry Briggs (1560-1631). Henry Briggs was a professor of mathematics at a college in London and admired the inventor of logarithms so much he took a year off and visited with Napier. The invention of Logarithms played the same role as the computer or calculator does today. It was a great short cut with regard to the chore of performing arithmetic operations. (See the above quote.)

Since the calculator and computer have replaced logarithms as a computing tool, you have a right to ask why we should study this concept. There are several answers. Henry Briggs, using the concept of logs, invented the slide rule. My guess is that some parent or your teacher still has a slide rule and could demonstrate how it was used. The reasons are:

1. It is part of the history of mathematics and the progress of civilization.
2. It will enable you to solve problems at this point which you can't solve or you would have difficulty solving with the calculator.
3. The logarithm function and its relationship to the exponential function is important in our world today.
4. It is a good mental exercise.

The first and fourth reasons above in all probability don't impress your student, but the second and third are creditable reasons as the applications will show.

Solving equations such as $5^{x+2} = 7^{x-2}$ or $1000 = 500(1.07)^x$ for x are quite easy using logs. This, hopefully, will create the mental motivation to encourage your student to look forward to this chapter.

Suggestion for reports

Computer Research: History
1. Who was John Napier?
2. Who was Henry Briggs
3. What are Napier's rods or bones? (Suggest you make a set and your student may want demonstrate how they were used to the class.)
4. How does the Slide Rule work? This engineers' calculator of the 1700 to early 1900's. (Your grandpa may have one.)

As you may guess, we will start with a definition.

Definition: If $n = b^L$, then the logarithm of n to the base b is L, and conversely. (b is positive, except in special cases.)

$Log_b n = L$ (Read as Log of n to the base b is L, which can also be expressed as $n = b^L$ or log of n is really an exponent.

Comment: You learned in geometry that all valid definitions are true when reversed. This is what conversely means, the statement reversed.

Is the following a valid definition? (no)

If you have a horse, then your animal has 4 legs. (The reverse is not always true. There are lots of horses in Montana.)

Examples:
1. $log_3 9 = 2$. How do we know? From the definition, the $log_3 9 = 2$ can also be written as $9 = 3^2$.
2. $2. log_{10} 100 = x$. Solve for x. $log_{10} 100 = x$ can also be written as $100 = 10^x$. You know from this expression that x is 2.
3. If $log_3 N = 3$, what is N? (The answer is 27. Can you explain why?)

Definition: If Log n with no base indicated, then we assume the base is 10.

This will take some effort and time on the parent's part for the student to be able to understand the definitions. Complete the following table will help.

Understanding Activity

Exponent form	Log form
1. $9 = 3^2$	$Log_3 9 = 2$

(Numbers 1 and 2 are completed. Write the Log form for each case below.)

2. $625 = 5^4$	$Log_5 625 = 4$
3. $64 = 4^3$	?
4. $8 = 2^3$	?
5. $32 = 2^5$	?

6. What is the log of 100?

(This type of question is usually on entrance exams.)

A few years ago students spent weeks practicing multiplying and dividing, taking roots and doing hard arithmetic operation using logs. Now the calculator and computer can do it much faster and efficiently. So why learn logs? There are problems that you need to understand and use logs to solve them.

Before the practice activity, one theorem should be proved just to re-enforce the true aspect of mathematics and the distinguishing factor which makes mathematics unique. So far, you have only worked with logs with base ten, which are in your student's calculator's memory. Can logs have other bases? The answer is yes and the following theorem will show you how to convert to other bases. (The base **e** is used is science. The number **e** is equal to the **limit of $(1+1/x)^x$** as x gets very large. Ask your student to guess what the limit is as x gets larger and larger. Then use your calculator to arrive at a value.) The letter e was selected in honor of L. Euler.

Theorem: If $Log_b N = X$, then $X = log N / Log b$.
  Proof: Given if $Log_b N = X$, which can be written as $N = b^X$. Why?

Now take the log of each side of $N = b^X$ which is $Log N = x log b$ (the base is now 10).

Your question, no doubt, is why or how is the log b^x equal to $x log b$?

The answer is that x is an exponent and log b is also an exponent and to simply exponent we multiply: $(10^3)^2 = 10^6$.

(Recall, If no base is indicated, it is understood to be Base 10.)

Now solve the equation (logn =xlogB) for X, therefore: $X = log N / log b$. Q.E.D. (What does QED mean? See a good dictionary.)

Examples:

1. Let's say you need to know the value of y in the following equation. (Your employer needs to know the value, but did not tell you why. Which employers will do.) Here is the equation, $3^Y = 12$.

    Your solution. You know that the answer is between what two integers? $3^Y = 12$, but you **must have the answer to 3 decimal places.** Do you understand that the value of y is between 2 and 3.

Logs will give a much more accurate answer. Carefully "walk" through with your student the following steps and be sure the student understand each one!
  a. Take the log of each side of the equation. $y\log 3 = \log 12$
  b. Divide by log3 which leaves y on the left side and log12/log3 on the right side.
  c. Using your calculator substitute the values for log12 and log3: Y = 1.07918/.47712
  d. Using your calculator to divide gives the answer y = 2.26186.

This agrees with your prediction that y is between 2 and 3. Your employer is happy now. What answer did you give the employer?

$$Y = 2.262 \quad \text{why?}$$

Being a good student, you checked the answer before giving it to the employer. The check is: $Y^{2.26186}$ equal to 12 or approximately?

2. Banking problem: How many years will it take $5000 to increase to $10000 at 6% interest compounded?

The problem stated in equation form is $10000 = 5000(1.06)^y$. (The banker's rule of 72 gives the answer of 12 years. Your student may recall the rule.)

Solution to the problem:

a. Divide by 5000 to simplify the equation.
$$2 = (1.06)^y$$
b. Take the log of each side.
$$\log 2 = y \text{ times } \log(1.06)$$
c. Substitute the values for the logs from your calculator.
$$.30102999 = y(.02530581)$$
d. Solve this equation using your calculator:
$$y = 11.895666 \text{ or } 11 \text{ years} + 10.7 \text{ months}$$

The banker would no doubt say 12 years since the problem stated compounded annually and he would round off the answer, or 72/6.
(The easy way is to recall the Banker's Rule of 72.)

Comment: The years needed is 11.895666 years. Notice the values were to 6 decimal places. The banker would no doubt say 12 years since the problem stated compounded annual.

Comment: Do a computer search for the life of Henry Briggs and John Napier. Perhaps your student would give a report to the class.

I would suggest the parent and student re-work these problems or similar ones several times over the next few weeks.

Now you need practice and more practice to really understand logarithms. Use the above examples or your notes as a guide to help you solve the following problems and the teacher assignments.

Questions

a. Is brushing your teeth and putting the paste on the brush a commutative operation?
b. Is reading a book and opening it a commutative operation?

Activity for Understanding

1. Use your calculator to find values for the following logs.
   a. log1000   b. log591.5   c. log68.48
   d. log3.186
2. Use your calculator to find the number N, which has the given log.
   a. logN = .08454 b. logN = 1.7577 c. logN = 2.5498

THIS SHOULD HELP YOUR STUDENT WHEN THE TEACHER EXPLAINS LOGS AND ASSIGNS THE HOMEWORK.

Practical applications that use logs

3. The time T for the period to complete oscillation of a Pendulum is given by the $T = 2\pi\sqrt{(L/g)}$, where L is the length of the pendulum in feet and g is the gravity constant or 32.16 feet per second per second. If a group of students made a 12-foot pendulum and hung it in a stairwell at the school, what would the time be for a complete oscillation or swing? First, use your calculator and then check your answer using logs. (Answer to the nearest second.)

    Answer: 3.68 seconds or 4 seconds

4. Another example problem is growth in nature. Let's assume the past growth in a town is 3% per year the last few years. The city council asked the math class to give them the estimation as to how many years it takes for the population to reach 32000, given the following information?

   A = 12000 population now
   The estimated population 32000 in Y years?
   r = 3% per year growth
   y = number of years

   Solve for y: $32000 = 12000(1.03)^y$
   Answer: 33 years (rounded)

   Question: How many years will it take for the population of your town to double if the growth rate is 3% per year? (Answer to the nearest hundredth.)

   Hint: $2 = 1(1.03)^y$
   Answer: 23.45 or 23 years 5 months

5. In a physics course you may have learned that sound is measured by decibels. One sound is a decibel louder than another sound when it is approximately 1.26 times as loud. (The Bel as a unit was named after Alexander Graham Bell. A decibel is 1/10 of a Bel.) Normal speech is rated at about 40 decibels, and injury to hearing begins at about 90.

   Comment: The comparison of two sound intensities is given by the formula $n = 10\log(I_2/I_1)$, where n is the ratio of the two sounds, I_2 is second intensity and I_1 is the first intensity. To compare the intensity of the noise at a rock

concert (120 decibels) to normal conversation we substitute the information in the formula as follows:
$$N = 10\log(I_2/I_1) \text{ and solve for } N.$$
$$N = 10\log(120/40) = ?$$
Explain your interpretation of the answer.
Answer: The rock concert is 4.8 (rounded) times as loud.

6. The most famous pendulum clock in the world is Big Bend in London. The pendulum is 13 feet in length.
    a. What is the period?
    $$T = 2\pi\sqrt{(L/32.16)} \quad \text{(See problem 3)}$$
    b. What is the length of a pendulum when the period is one second?

Answers:
a. 3.99 seconds or 4 seconds,
b. Approximately .81 ft or 9.8 inches.

Interesting Application: Sound

7. Which medium will transmit sound travel better, air or a solid wall? Select your answer, then conduct the following experiment with a friend.
    a. Stand at one end of a room and your friend at the other end of the same wall. Tap the wall with your fingernail or a pencil. Can your friend hear the tapping?
    b. Now have your friend put their ear to the wall and you tap the wall again. Can the other person hear the tapping? Repeat the experiment until a conclusion is reached. (You may have to increase the Hardness of the tap.)

**The denser material will conduct sound better.
Note: Sound travels about 13 times faster in water then in air.**

Thinking Activity: Easy Money

Your student and friend are having a coke after school in the lounge. The friend puts a half dollar on the table and you do the same. The same process is repeated two more times. How much is now on the table? Your friend now offers you the coins on the table if you will give her $2. Should you accept? Is this a "good" deal?

THINK
Explain or justify your answer.

Answer: Do not accept the "good" deal. Act it out and your student will see the why?

College Student

A college student sends an email to his parents and requests some money. The message was brief and questionable as to the amount. But the father knew the son liked math and interesting problems.

The message is below and the father assume the letters represented integers and it was addition and he put the decimal points in. Each letter represents only one number. In other words, if the E's are 5, then no other letter can be 5.

<pre>
        SE.ND   What number is M and why?
      + MO.RE   M + S must be greater than?
        MON.EY   What amount did the
                 students request?
</pre>

Note: There are several possible answers! Give your student a treat when he or she solves it.

Chapter 8. The High School Years - Grades 9-12
Section 8f. Trigonometry for Right Triangles

TO MEASURE IS TO KNOW

Johannes Kepler

This chapter extends over a period of 2500 years and is still very useful. The first applications were solving for unmeasurable distances, involving problems in construction and in navigation. Today the problems are surveying, navigation, construction and other fields. The playing field today, is not only on land and sea, but also in space, plus the more recent applications in the medical world. It is more important for a person to have an academic foundation in Mathematics today, then ever before. Remind your student to always ask questions.

Parent: Keep in mind this material is to help you and your student to understand what the teacher is explaining and assigning as homework.

As stated above, this material extends over a period of 2500 years and is still very useful. The first applications were solving for unmeasurable distances, involving problems in construction and in navigation. Today the problems are still surveying, navigation, construction and many other areas. The playing field today, is not only on land and sea, but also in space, plus the more recent applications in the medical world. It is more important for your student to have an academic foundation in mathematics today than ever before. An "academic foundation" means an appreciation of mathematics and its contributions. It does not mean the student can do all the mathematical operations. Keep in mind you are just touching the tip of the mathematical "iceberg."

Another quote pertaining to mathematics is, **"To measure the unmeasurable."** How can you measure the unmeasurable? From that question alone, this chapter should be interesting or at least arouse your curiosity. Can you think of a few objects that would fit into the unmeasurable list? Hint: The distance around the equator. The ancient Greeks had many similar questions just like you may have. This chapter begins with some concepts, which have their origins back in the days of the Golden Age of Greek Mathematics. (See "Golden Age" in Boyer's book: *A HISTORY OF MATHEMATICS.)*

Research: Trigonometry, including information as to:
    a. The history of Trigonometry with regard to when, where, and why.

    b. Who was Hipparchus?

Suggested references:
    GREAT MOMENTS IN MATHEMATICS BEFORE 1650, Eves, H.

    THE MATHEMATICS AND PHYSICAL WORLD, Kline, M.

    Or any book on the History of Mathematics

The Trigonometric Functions
Sine, Cosine, and Tangent Functions

We have had several terms, which may be unknown to you in just the first few paragraphs in this chapter. Can you identify the terms? The first word needing explaining or defining is used in the title for the chapter, Trigonometry. What do you see in the word, trigonometry? Your first answer is probably "metry", which is associated with measure. The second word you would pick out is "tri", which is associated with three as in triangle. Most students then ask where the middle part "gono" fits in. The answer is the Greeks actually used the word trigon or trigonon, which means, triangular in shape. Putting these two together we have triangle measure or trigonometry. (What is etymology?) The meaning of the terms SINE and COSINE will be given at the appropriate point.

How did the Greeks develop this branch of mathematics and why? The why may be easier to explain then the how. We have weather forecasters today and you probably hear their predictions every morning on radio or TV. In the ancient days, these forecasters predicted the times for planting, for harvesting, and the days for what we call holidays. We also have land surveyors who are usually checking the boundaries for a lot or determining boundaries or key points for a freeway or toll road. The ancient surveyors determined the boundaries after a flood to determine which land belonged to the king, etc. One of their tools was a piece of rope marked off in 3-4-5 segments. Can you figure out why? (Hint: Pythagorean Theorem.) The surveyors of the early times were also called rope stretchers. Their tool was as indicated above (the 3 by 4 by 5 rope) which when stretched at the vertices of the 3-4-5 right triangle formed a right triangle. It was a convenient way to determine a corner for a lot. Suggest your student make this tool and demonstrate the method to the class. Today, of course, modern surveying instruments are used.

Comment for teacher: Discuss surveying with a local surveyor or invite him or her to class and explain their tools and methods. The new methods are very interesting.

If two right triangles are similar, then the ratio of the side opposite an acute angle divided by the hypotenuse is a constant. (Ask your student the conditions for similar triangles.) The Greeks named this ratio the SINE of the angle and we still use the term today.

Comment: The term sine came from the Hindu with reference to the chord of a hunter's bow. This term will be further explained.

Definition: The SINE (SIN) of an acute angle in a right triangle is the ratio of the length of the side opposite the angle divided by the length of the hypotenuse.

Sin A = a/c

Sin B = b/c

Fortunately, these ratios have been worked out and are in your calculator. Find the sin button on your calculator. In order to find the sin of the acute angle your calculator must be in the degree mode. (If you don't have your book of instructions that came with your calculator, then ask your teacher or parent to help you tell when your calculator is in degree mode.)

Example: What is the sin of 37°?

Steps	Display
1. Turn calculator on.	0
2. Press key 3 then 7.	37
3. Press sin key.	.6018 (4 places)

Comment: If your student did not get that answer, check the instruction book that came with your calculator, or ask the teacher. Calculator must be in the degree mode!

Activity for Understanding
(Use your calculator)

1. What is the sin of the following angles? (Angles are in degrees)
   a. 10 b. 13 c. 30   d. 45   e. 55   f. 60
   g. 75 h. 82 i. What do you observe as to the value the sin function as the angle increases?
   Answers:
   a. .173648 b. .224951 c. .500000 d. .07107
   e. .819152 f. .866025 g. .965926 h. .90268

2. This time the problem will be reversed. Using your calculator, record the measure of the angles whose sines are listed below. (Answer to the nearest degree.)

Comment: You may need to check the instruction book for help with this inverse operation. What is the size of the angle if given?

   a. $\sin A = .42266$ b. $\sin A = .6018$
   c. $\sin A = .88295$ d. $\sin A = .60656$
   e. $\sin A = .95630$ f. $\sin A = .20791$

   Answers: a. $25°$ b. $37°$ c. $62°$ d. $41°$ e. $73°$
   f. $12°$

The COSINE Function
(Pronounce co-sign)

In the figure below the method used for labeling angles and sides is capital letters for vertices and small letters for the sides. The small "a" represents the side opposite the vertex A; likewise, the small "b" is opposite vertex B. This is nothing new to you since this method was used consistently in your geometry course and possibly other previous courses.

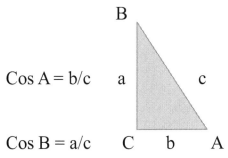

Cos A = b/c

Cos B = a/c

The COSINE (abbreviated COS) of an acute angle in a right triangle is the ratio of the measure of the adjacent side divided by the measure of the hypotenuse.

Definition: The COSINE of an acute angle in a right triangle is the ratio of the length of the adjacent side divided by the length of the hypotenuse.

Use your calculator to find the Cos of 43°

Steps	**Display**
1. Turn calculator on.	0
2. Press buttons 4 then 3.	43
3. Press cos button	.7314 (4 places)

Comment: If your student did not get that answer, help the student and try again.

Activity for Understanding
(Use your calculator)

1. What is the cos value for the following angles? (The measure of the angles is stated in degrees.)
   a. 17 b. 26 c. 30 d. 45 e. 56 f. 60 g. 72
   h. 87 i. 89 j. As the measure of the angle increases, what does the value of the cos approach?

Answers:
a. .956304 b. .898794 c. .866025 d. .707106
e. .559193. f. .5 g. .309017 h. .052336
i. .001745 j. The value approaches 0.

2. What is the degree measure of the angle for the following?
   a. Cos A = .29237 b. Cos B = .48481
   c. Cos A = .91355 d. Cos A = .54464
   e. Cos B = .97437 f. Cos A = .08716

Answers are in degrees rounded to the nearest degree:
   a. 73 b. 61 c. 24 d. 57 e. 13 f. 5

Now your student is possibly beginning to question, "For what are the sin or cos functions used?" Here are two examples and then some practice problems. In the following exercise, a student is given the problem to determine the length of the side indicated by a, using the given information indicated.

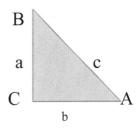

Given: m ∠ A = 40°, hypotenuse = 110 ft., then side a = ?

Solution: The side the student needs to solve for is opposite the given angle. This means you need to use the sin function, since the side opposite is the side in question and the hypotenuse is known. (See definition.)

Writing the equation: sin 40° = a/110
Substituting the value of sin 40° from your calculator and then solving for the unknown results, a is equal to: 70.706637 feet.

.6427888 = a/110
Solving for a by algebra:
110(.642788) = a = 70.706637 ft.

Do you think this is a good answer? Why all the decimal places? We will agree to always give the answer to fit the original data! What do you think the 110 ft. was measured with? Let's assume a measuring device like a tape measure and the measurement is to the nearest foot, since the 110 doesn't have inches or a decimal fraction indicated. Therefore, we will give our answer to the nearest foot, or in this case 71 ft.

Another example

A ramp is 21ft long and 18.5ft. from A to C as indicated in the figure on next page. What is the angle of the incline (angle A)? For safety regulations, the angle cannot be > than 30°.

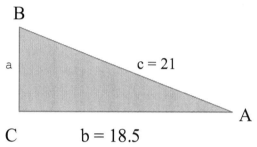

From the figure, the following equation can be written: cos A = 18.5/21

Dividing 18.5 by 21 with your calculator, the display reads .88095(rounded).

Now we need to know the angle that has a cos equal to .88095.
Press inverse or second function button on your calculator, then the cos button and the display should read 28.24 (depending on your type of calculator) or rounded to 28 degrees and 14 minutes. Where did the 14 minutes come from?

Comment: The method you use in rounding answers is important. The answers should always make sense considering the data in the problem. The answer should have at least the same degree of accuracy as the values in the problem.

Activity for student understanding

In working the following activity, keep in mind the trigonometric functions at this point only work with right triangles. In other words, to use the sin or the cos functions you must have a right triangle. If the right triangle isn't given you most create one in the figure by your ingenuity. Draw a figure for each problem.

1. Calculator problem: What number does the value of the sin appear to approach as the angle gets larger in problem 1? As the angle approaches 0?
   Answer: 0, 1

2. What number does the value of the cos appear to approach as the angle approaches 90? As the angle approaches 0?
   Answer: 1, 0

3. What are the Sin and Cos of the acute angles in an isosceles right triangle?
   Answer: 7071 to 4 places

4. Convert the following to degrees and minutes:
   a. 16.4 degrees is equal to 16 degrees and ? minutes. Hint: .4 of 60 equals ?
   b. 61.7 degrees is equal to 61 degrees and how many minutes?
   c. 57.5 degrees is equal to 57 degrees and how many minutes?

d. 71.12 degrees is equal to 71 degrees and how many minutes?

Answers: a. 24' b. 42' c. 30' d. 7'

5. a. Do you think the sin of 25 degrees is 1/2 the sin of 50 degrees? Check your answer using your calculator.
    b. The sin of 25 degrees is 1/2 the sin of what angle?

Answer: a. Your answer   b. 57.7 degrees

The Tangent Function

In the figure below the method used for labeling angles and sides is capital letters for points and small letters for the sides. The small "a" represents the side opposite the vertex A, likewise the small "b" is opposite vertex B. This is nothing new to you since this method was used in your geometry course and possibly other previous courses. See the figure below.

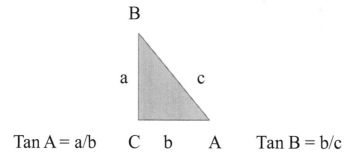

Tan A = a/b   C   b   A   Tan B = b/c

The Tangent (abbreviated Tan) of an acute angle in a right triangle is the ratio of the measure of the opposite side to the measure of the adjacent side.

Definition: The Tangent of an acute angle in a right triangle is the ratio of the length of the opposite side to the length of the adjacent side.

Use your calculator to find the Tan of 43°

Steps	Display
1. Turn calculator on.	0
2. Press buttons 4 then 3.	43
3. Press Tan key	.93252 (5 places)

Comment: If your student did not get that answer, help the student to try again.

Activity for student Understanding

1. What is the Tan value for the following angles? (The measure of the angles is stated in degrees.)
   a. 17 b. 26 c. 30 d. 45 e. 56 f. 60
   g. As the measure of the angle increases, what does the value of the cos approach?

   Answers: a. .3057 b. .4877 c. .5773 d. 1
   e. 1.4826 f. 3.0777 g. The value approaches 0.

2. What is the degree measure of the angle for the following?
   a. Tan A = .29237 b. Tan A = .48481 c. Tan A = .91355 d. Tan A = .54464 e. Tan A = .97437 f. Tan A = .087165

   Answers are in degrees rounded to the nearest degree. a. 17 b. 26 c. 42 d. 29 e. 44 f. 5

Now your student is possibly beginning to think, "What is the Tan function used for?" Two examples and then some practice problems.

In the following exercise, a student is given the problem to determine the length of the side indicated by **a**, using the given information indicated.

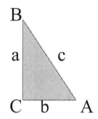

Given: $m\angle A = 40°$, side $a = ?$, $b = 110$ ft.
$\tan 40° = a/110 \rightarrow a = 112(\tan 40°)$ or $a = 92.30$, therefore a is 92 ft.

The student is probably wondering what to do if the triangle is not a right triangle? Good question!

CHAPTER 8. The High School Years - Grades 9-12
Section 8g. Trig. for Non-Right Triangles

In the prior section, you learned how to find the sin, cos, or tan of any positive angle. The method can be adapted to work with negative angles. In this section, you will be briefed on how to use these functions to solve problems involving triangles, with no limitation as to type of triangle. In other words, the triangle doesn't have to be a right triangle. Many of the problems in the real world involve right triangles but many do not, such as land surveying, road construction, and navigation on the land, sea, and in air or space. Trig also plays a vital role of power factor connected with electrical energy transmission as well as many medical related applications.

The solutions to non-right triangle problems involve two theorems, the Sin Theorem and the Cos Theorem. These are useful theorems since they work for right triangles as well non-right triangles. If you know these two theorems and can apply them, you can solve many applications involved in architecture, navigation, medical, military, surveying, etc. Since this is not a text book, but the objective is to only assist the parents and/or teacher to help the student to better understand what is being taught in order to, hopefully, insure success.

The following two theorems will be justified, even though the school's text may not prove them.

The Sin Theorem

The name of the inventor of the sin theorem has been lost in history so a report can't be recommended pertaining to that person and his or her life. The following problem is one that the ancient mathematician tried to solve. In the figure below, let's assume the point S is a ship, and the points A and B are on the shore 10 kilometers apart.
"TO MEASURE THE UNMEASURABLE"

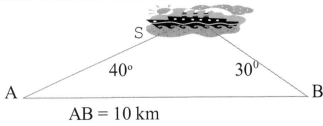

181

This could be a military or a rescue problem where S is the ship and the defensive guns or rescue teams are at A and B. Distances AS and BS are required.

The "guns" may have been rock throwers or catapults in ancient times and rockets today.

Now that you are hopefully curious as to how the problem is solved, the method will be explained. First, the general solution will be derived. Follow these steps, ask questions where explanations are needed.

Convert the problem to right triangles by drawing the perpendicular or plumb line CD. This is done so you can use the theorems for the right triangle from the previous sections. (Recall it was stated in one of the previous sections that we try to convert a new problem to an old problem that we know how to solve.)

Step 1. Draw a triangle similar to the one below, and label the figure as shown.

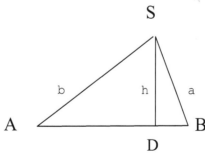

Step 2. Convert the problem to right triangles by drawing the perpendicular or plumb line SD. This is done so you can use

the theorems for the right triangle from the previous sections.

Triangles ADS and BDS are right triangles.

Step 3. Notice side AS is labeled b, side BS is a, and SD will be h to make communication easier. Do you agree with the following?
    a. Sin A = h/b   b. Sin B = h/a

Step 4. If Sin A = h/b, then h = b(sin A) and if Sin B = h/a, then h = a(sin B)

Step 5. Since h = h, and by substituting from step 4 you have: h = h or b(sin A) = a(sin B) and dividing by ab gives:

$$\frac{\sin A}{a} = \frac{\sin B}{b}$$

Consequently, the following theorem can be stated. Notice we almost proved the theorem, but left the Sin C/c case for the parent and student or the teacher to justify.

Sine Theorem: If given a triangle, labeled ABC, then
$$\frac{\sin A}{a} = \frac{\sin B}{b} = \frac{\sin C}{c}$$

Do you see why it is called the Sine Theorem? It should be easy to remember! Now back to our problem to determine the distance the ship is from A or B.

Draw the figure and label it to fit the following information. We know that AB is 10 kilometers, angle SAB is 40 degrees and angle SBA is 30 degrees.

What is the measure of angle ASB?

Answer: 110 degrees.

Write the equations using the sin theorem.

$$\frac{\sin A}{a} = \frac{\sin B}{b} = \frac{\sin S}{s}$$

Substituting the information from the problem, you have:

$$\frac{\sin 40}{a} = \frac{\sin 30}{b} = \frac{\sin 110}{10}$$

Using: $\frac{\sin 110}{10} = \frac{\sin 40}{a}$ and solving for a:

therefore, $a = \frac{(\sin 40)10}{\sin 110} = \frac{.6428(10)}{.9397} = 6.84$

BC or side labeled a equals 6.84 kilometers or approximately 4.25 miles.

In like manner the distance side labeled b can be determined using:

$$\frac{\sin 30}{b} = \frac{\sin 110}{10}$$

Substituting

$$\frac{\sin 30}{b} = \frac{\sin 110}{10}$$

therefore, AS or $b = \frac{(\sin 30)10}{\sin 110} = \frac{.5(10)}{.9397} = 5.32$ km.

184

This means the distance from A to S is 6.84 km which is r approximately how many miles? (use the I table on page 124 or Index 5) The final answers are rounded and SB is 5.32 km or how many miles? You must admit these ancient mathematicians were very clever. This also makes it evident why students study algebra before they study geometry.

Activities for Understanding the Theorems

The following procedure or steps will help your student solve the problems.
  a. Draw the figure. b. Label the figure.
  c. Identify the unknown. d. Write the equations.
  e. Solve for the unknown. f. Answer the questions.
(Organize your work so it is easy to follow your logic to the answer.)

1. An engineer for a road construction company is given the problem of calculating the length of a proposed new section of a bike path. (Indicated on the map below as BC) It is known that ∠A is 42°, ∠C is 56°, and AB is 3457 yds. Originally a person from B had to ride to A and then to C. The new path (BC) will be faster, shorter and more scenic. (Figure below is not drawn to scale. What are the unknown values on the figure?)

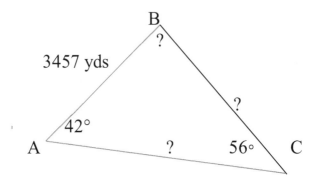

Answer: BC = 2790 yds, angle B=82 °   AC =4130

2. In the following triangle, solve for the two unknown angles and the unknown side. (Not drawn to scale.)

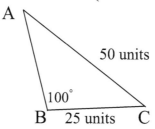

Answers: A = 29.5 degrees, C= 50.5 degrees, 39.2

3. The following diagram illustrates a problem a cable television company had. Instead of running cable from A to B and then to C, the CEO asked their engineer how many feet of cable would be saved if they ran the cable directly to C from A? The accountant knew it was 1200 ft (AB=500 and BC=700) from A to B to C and at $7.50 per foot the cost would be $9000. The engineer said: "Give me a few minutes with my calculator and using the sin theorem I'll have an answer for the length of AC."

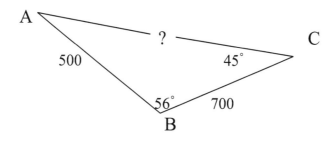

a. Length of AC is between which two integers?
b. What is the length of AC?

c. What is the cost for cable from A to C?
d. What is the dollar savings?
e. Do you think the CEO was impressed?
   Answers: a. 200 < ? < 1200   b. 586   c. $4395
   d. $4605   e. yes

Challenge

A cube has a side of 2 inches. Draw the 3D view of the figure.
   a. What is the measure of the diagonal of the base?
   b. What is the measure of the diagonal of the cube?

   Answers: a. $2\sqrt{2}$ b. $2\sqrt{3}$

The Cosine Theorem

One day after thinking about the Sin Theorem, a student asked the professor, "How do you solve a non-right triangle problem if all that is given are the three sides, or two sides and the included angle?" Very good question, since the Sin Theorem appears not to handle this case due to the fact that in order to use the Sin Theorem you must know an angle and the opposite side plus one more fact, a side or an angle. The professor said: "Use the Cosine Theorem!"

The following problem will help you understand the Cosine Theorem.

Given: Triangle ABC with AC = 5, BC = 7, and AB = 9

Question: What is the measure of the angles to the nearest

187 degree?

Draw the figure: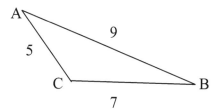

In order to solve this problem, the professor showed the student a more general problem and informed the student, that if the general problem is solved, then we just need to adapt the specific problem to the general formula and substitute the values into the formula. This is a nice proof for the teacher to justify in a class period. It takes time but it is worth it. (Capital letters are points.)

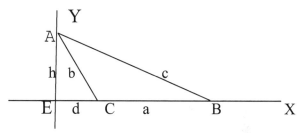

Step 1.: $c^2 = h^2 + (d + a)^2$ or $c^2 = h^2 + a^2 + 2ad + d^2$
Step 2. But $b^2 = h^2 + d^2$ or $h^2 = b^2 - d^2$
now substitute in step 1, $b^2 - d^2$ for h^2
or $c^2 = b^2 - d^2 + a^2 + 2ad + d^2$ or
$c^2 = b^2 + a^2 + 2ad$.
Step 3. Therefore:
$c^2 = a^2 + b^2 + 2ab\{\cos \text{angle}(ACE)\}$

Notice Angles ACB and ACE are supplementary since we know the angle ACE but we need angle ACB. Let look at the two angles and see if the cos values are related. If cosine angle ACD (10°) is .9848 what is the value of cos of angle ACB (170°) – .9848. complete the following table:

Angle ECA	cos	Angle ACB	cos
20	.9397	170	– .9397
30	.8660	150	– .8660
101	.1908	79	– .1908

Conclusion. In place of angle ACD we can substitute – cos of ACB.

Cos Theorem: If given the sides of a triangle are a, b, and c, therefore the Cos theorem is:
$$c^2 = a^2 + b^2 - 2ab(\cos C)$$
or
$$a^2 = b^2 + c^2 - 2bc(\cos A)$$
or
$$b^2 = a^2 + c^2 - 2ac(\cos B)$$
Now your student can solve all the Trig Cases!

Given the three sides, help your student convert the formula to this form and solve for the angle:

$$\cos C = (c^2 - a^2 - b^2)/ - 2ab$$

A special Problem (A must!):
Solving for the value of Pi (π)

Definition: Pi is the ratio of the circumference to the diameter.

189

The perimeter of the inscribed polygon in the circle approaches the circumference of the circle as the number of sides of the polygon increases.

Given: Chord AB with midpoint D. Radius of the circle is 1. OD is perpendicular bisector of chord AB. Draw the figure on your paper.

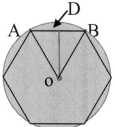

Figure: Right triangle AOD with angles of 30, 60 and 90 degrees. (Students may need to justify the angles sizes.)

How many right triangles like ADO can be drawn in your figure?   Answer is 12. What is the measure of angle AOB and angle AOD? (60,30)

Solution: Sin of angle AOD is AD/AO and as the number of sides of the inscribed polygon **increases** to n sides and R, the radius is 1, then angle AOD is 360/2n or 180/n and side AD is R(sin180/n). The Perimeter is 2n(AD) but AD = R sin 180/n, therefore the Perimeter is the sum of the ADs or n(2AD) and when you divide by AO (which is 1) you have 2n(sin 180/n) substituting for AD, the perimeter equals 2n(R)(sin180/n) or the perimeter **approaches** the circumference. Since Pi is C/D and the perimeter is very close to the circumference **when n is very large,** we can write:

$$P/2R \text{ approaches } C/2R = Pi$$
$$Pi = \frac{2n(R)(\sin 180/n)}{2R} = n(\sin 180/n)$$

a. Calculate the value of Pi using the above formula for the following values of N.(Record your answers.) 50, 100, 250, 500, 750, 1000, 1500. 2000

b. Write your conclusion.

Update your Notes and Comments

Research: How did Eratosthenes calculate the circumference of the earth about 250 years BCE? (See book by L. Hogben.)

Notes

In Missoula, Montana, we have a fire fighting training facility. If you are ever here it is worth a visit. Fire fighters and Forest Rangers dislike a slope of 1. What does it mean? Check your calculator for the angle with a Tangent of 1. (the slope is 45 degrees.)

**Chapter 8. The High School Years, grades 9-12
Section 8h. THE NORMAL CURVE**

Statistics makes possible new perceptions and realities by making visible large-scale patterns.

Neil Postman

Your student has no doubt heard of the term "normal curve" and/or has even asked the teacher the question, "Are you grading on the curve?" Did your student really understand the meaning of the term and the consequences of the question if the answer was yes or no? This section will help the student understand the meaning and the derivation and use of the term "normal curve." It is used, very often, in business, education and the sciences.

The application of the term normal curve was really a by-product of World War I. The curve was first introduced by Abraham De Moivre (1667-1754) and Karl Gauss (1777-1855). Sometimes it is called the Gaussian Curve. The U.S. Army (World War I) began to draft men and naturally had to provide the recruits with wearing apparel such as shoes. One procedure would be to measure the size of each soldier's feet and then order the shoes. Naturally, this would take time and create an awkward situation. Another option would be to order large quantities of shoes in all possible sizes. This would probably leave a number of pairs not used like the very small sizes or very large sizes. The solution was to measure the sizes of a large random sample of men, and then assume the new recruits would fit the same pattern of frequency for sizes and hence, order the shoes in advance of the need. This is what was done, and the curve resembled the one below.

Suggested Research: Karl Gauss and Abraham De Moivre

From the curve, an estimate can be derived as to the number of pairs to order of each size. You can see more size 9 would be ordered than any other size. (This is over simplified!)

Suggested Research: Karl Gauss and Abraham DeMoivre

This procedure was also applied to other equipment needs where size is a factor.

The NORMAL CURVE is also known as the BELL-SHAPED curve. The name "bell shaped" is due to the shape of the curve. It is also identified as the "natural curve" due to the fact that many patterns in nature resemble the curve. This curve has several important properties.

Definitions: Properties of the Normal or Bell Shaped Curve:
1) The Mean, Mode, and Median are all equal or the same value and are on the line of symmetry.
2) Plus and minus one STANDARD DEVIATION (determines the amount of spread on each side of the mean or line of symmetry) will include approximately 68% of the data.
3) The STANDARD DEVIATIONS on each side of the mean (line of symmetry) will include approximately 95% of the data.
4) Three STANDARD DEVIATIONS on each side of the mean (line of symmetry) will include approximately 99.8% of the data.
5) 5 From the data or the curve the RANGE (The measure from the lowest to the highest) can also be determined.

The above is summarized on the graph below using the shoe sizes from above. The black segments indicate the Standard Deviations.

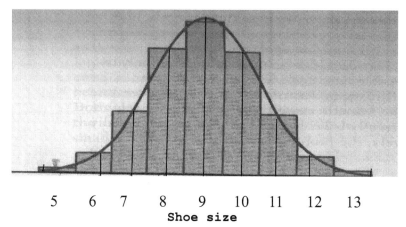

Your question no doubt is: How is the value for STANDARD DEVIATION calculated? What we need are some definitions!

Definition: The RANGE is the difference between the highest or largest number and the lowest or smallest number in the set of data. (In the shoe case above, the range is 13-5 or 8.)

Definition: The Standard Deviation of a set of data is the square root of the mean of the squares of the deviations from the mean (the black segments).

You may need to read the definitions a few more times in order to really understand them. The following formula is the mathematical expression of the definition.

$$SD = \sqrt{\frac{(s_1-m)^2 + (s_2-m)^2 + (s_3-m)^2 + ... + (s_n-m)^2}{N}}$$

Where **m** is the mean score, **s** is each individual score or set, and **N** is the number of scores. You can see this takes some effort and your calculator is definitely needed. **Some**

calculators have a built-in program for calculating the standard deviation. Check the instruction book for your calculator.

You also, no doubt, wonder where the percentages, (68%, 95%, 99.8%) came from (see definition). These are derived by using CALCULUS, a college course in mathematics and the numbers refers to the area under the curve. This means you will have to wait until you take the calculus for the explanation and just accept it for now. (Call it an assumption if you wish.)

The following example will help you understand the Standard Deviation concept, provide an appreciation for the effort to calculate this information, and how it is used in interpreting the "picture." The following data are the results of surveying 100 male students as to their shoe size

Size	Frequency	Deviation from mean	Squared	Squared x freq. Total	
6	2	3	9	18 (2x9)	
7	12	2	4	48	N= 100
8	19	1	1	19	
9	28	0	0	0	M = 9.1
10	21	1	1	21	
11	15	2	4	60	SD = 1.4
12	3	3	9	27	

1. a. The first chore is to calculate the three averages (mean, mode, and median).
    (1)Answer: The mean is 9.1 rounded to 9.

(2) Mean = mode = median = 9 (rounded) and the data is assumed to be normal.
   b. Now using S.D. definition to arrive at the Standard Deviation or SD, it turns out to be the $\sqrt{1.93}$ or 1.4 rounded.

$$SD = \sqrt{\frac{(s_1-m)^2 + (s_2-m)^2 + (s_3-m)^2 + ... + (s_n-m)^2}{N}}$$

   This tells us that 68% of the men should take sizes 8,9, or 10. About 14% will take size 7 and the same number for size 11, with about 2% taking size 6 or 12. My guess is if you were the buyer you would order more 9's then any other size.
2. Now draw the graph for the data and include the line segments showing:
   a. the mean, mode, median.
   b. + and - 1 SD.
   c. + and - 2 SD.
   d. + and - 3 SD.
   e. Include a title and label the axes.

Some person once said, "Mathematics is not a spectator sport." This means in order to understand the concepts and learn the skills one needs to practice (homework)!

Activity for Understanding (Team Problem)

Show your organized work on this team work (parent and student) problem!

1. Calculate the range, mean, mode, median and Standard Deviation (nearest tenth) for the following set of data. (If your calculator is programmed to calculate the Standard Deviations, use it for a check.)
   Data: 6, 7, 12, 14, 15, 18
   Answers: a. Range = 12, M = 9 ,Mode = 9, Median = 9, SD = 1.4
2. Your state has given a graduation test to 407,286 high school seniors over the past ten years and established the following curve that is very close to the normal curve.
   1 SD = 68% 2 SD = 95% 3 SD = 98% 4SD 100%

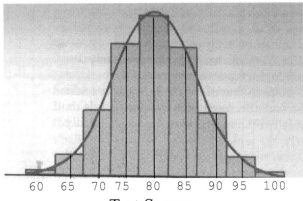
Test Scores

The mean is 80 and the Standard Deviation is 5.(See the definition for meaning)
   b. What is the ACTUAL percentage of students scoring higher than 80?
   c. What should the percentage be according to the Definition?
   d. What is the percentage of students scoring higher than70?
   e. What is the percentage of students scoring higher than 85?

  f. What is the percentage of students scoring higher than the percentage of students scoring between 75 and 85?
  g. What is the percentage of students scoring below 65?

  Answers a. 50%   b. 68%   c. 97.7%   d. 15.9%
  e. 68.2%   f. .1%

3. If two classes on the same test have the same mean but the SD of one is 5 and the SD of the other is 7, what does this tell you about the curves?
Write your conclusion and draw the possible curves on the same set of axes.

  Answer: The test with an SD of 7 has a wider range or spread.

Chapter 8. The High School Years – Grades 9-12
Section 8i. Types of Statistical Graphs

There is a saying: "A picture is worth a thousand words." What this section will do is help you learn or review different ways to represent a set of data. In section 1 of chapter 5, your student studied the three averages known as measures of Central Tendency. You student also discovered that three averages do not tell the whole story. In fact, they can be misleading and possibly lead to false conclusions. To prevent false conclusions a picture or graph method may be required.

There are several types of graphs to represent data, but in each the axes must be properly labeled, a correct title used, and the graphs must be easily interpreted. Many times color is used to provide a "colorful" aid for interpretation and attention.

Some of the types of graphs are reviewed below by using a small set of data for easy interpretation. The same data is used for the various types of graphs. A true picture takes a large set of data to tell if it approaches a normal curve.

Science Test Scores	Number of students (Frequency)
100	1
90	3
80	5
70	4
60	1

Type 1. Bar graph   Type 2. Line graph

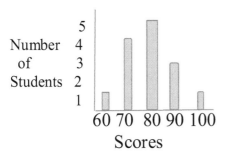
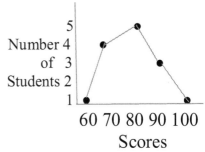

Type 2 (above) graph can lead viewers to a false conclusion that there are scores in between the numbers.

Type 3. Circle graph (Team project)

SCIENCES TEST SCORES

Help your student construct the circle graph. There are 14 test results, so 1/14 of 360 is 26 or 26° degrees for each score therefore, the degree for each group is:

Test Score	Number of students	
100	1	portion of circle is 26^0
90	3	portion of circle is 77^0
80	5	portion of circle is 130^0
70	4	portion of circle is 104^0
60	1	portion of circle is 26^0

What is the total degrees? (Shouldn't it total 360°?)
(Explain why the total is more than 360! (Rounding)

Now help your student construct the graph. This also brings in how to use a protractor or estimating.

Type 4. Three-dimensional types (Which could be done for all of the types.)

Suggest you use a type of 3D figures related to the subject, like the one below for boxes.

Type 5. This type uses meaningful symbols, such as letters, boxes or figures related to the topic.

Comment: To be meaningful each graph must have sufficient information on it to enable a correct interpretation, or to "tell the correct story!" Ask your student which type is preferred and why?

Comment: The circle type seems to be the hardest for most students to construct due to the use of degrees.

Let your student cut out from a newspaper or magazine a few graphs, and discuss the interpretation and point out any weaknesses or errors. This makes for an interesting parent-students team investigation.

The Center of Gravity Problem
Team work Investigation

1. Draw a triangle on a fairly stiff piece of cardboard, use your ruler and construct the three medians, then mark the point of intersection as the center of gravity. Cut out the triangular area and with a pin observe if the point where the medians intersect is the center of gravity by attempting to balance the triangular region on the point of the pin.
2. Draw a rectangle on a piece of cardboard, cut it out, then use your ruler and draw the diagonals. Attempt to locate the center of gravity using the pin or pencil. Is it the point where the diagonals intersect?

3. Locate the center of gravity for a ruler by the balancing method.
4. Locate the center of gravity point on a baseball bat by balancing.*

*Comment: The center of gravity is the point where the most energy is transferred to the ball, or the point where the batter hopes the bat will make contact with the ball. It is also called the sweet spot. (Ask your coach to explain. Hint: Use your finger tip as the balance point.)

Extra Think Problems

1. Ask the driver training teacher when a car will tip over, relative to the center of gravity point?
2. Do you see a pattern in this set of numbers pattern?
   1+2+3+5+ 8+13+21+?+?
   What are the missing integers?

Chapter 8. High School Years – Grades 9-12
Section 8j. Casino Math Probability

The mathematical meaning of the word probable

This section will introduce your student to the world of probabilities and help you understand mathematical probability compared to empirical probabilities. These activities will provide practice in calculating simple probabilities.

Students find this section the most interesting and practical since they frequently hear references involving probabilities in the news media. Many states today have lotteries as a means of raising revenue to cover the cost of government. Games, ancient and modern, involve activities incorporating the element of chance. Games like the toss of a die, the spinning of a wheel or the draw of a card are still very popular. Probabilities are involved in all forms of health and life insurance, as well as the life span of many objects your parents buy.

Certainly, it is used in weather forecasting, which you no doubt hear every day, and of course the references to it in everyday conversation. The quote at the beginning of this section states it all inclusively. The following quote will become meaningful as you complete this section.

The house plays the percentages (probabilities) while the player relies on luck.

H. Gross & F. Miller
(see bibliography)

The term mathematical probability is used, since it is based on the calculated way an event should happen, which may be very different from the actual results. But first, let us carefully define the terms so we all agree, thus avoiding misinterpretations in the future.

Definition: Probability of an outcome is the ratio of the number of favorable outcomes or successes to (or divided by) the total number of possible outcomes.

This in formula form can be simply stated as **P = s/n**, where **s** represents the number of successes and **n** the number of possible outcomes.

If your student is attempting to determine the probability of tossing a 4 with a die, the student may toss the die 10 times and counts the number of times a 4 turns up. Let's assume one 4 is turned up. The student could predict the probability for a 4 is 1/10 or 10% of the tosses. This method of determining the frequency of an event is called empirical probability.

This illustrates what mathematicians call **empirical** probability - probability based on actual trials. If the number of trials is very large, the results should approach the **theoretical** mathematical probability. In the above case, the mathematical probability is 1/6, since there are 6 ways a die can land when tossed and one of these is a 4. The empirical probability of an event should approach the theoretical probability as the cases increase and **the game is fair.**

Definition: Empirical probability is the probability based on actual trials.

So, you can see there are two kinds of probability - theoretical and empirical. In an ideal situation, the two will be the same ratio. Don't bet on this being the case. The empirical probability is very seldom the same as the theoretical or mathematical probability.

Understanding Activity for Parent and Student

1. Complete the following table for tossing a dye. The number of times a three turned up is listed.

Number of Tosses	Total Tosses	Successes For a 3	Total 3's	Empirical Probability for a 3
20	20	2	2	1/10
30	50	6	8	8/50
50	100	8	16	16/100
70	170	12	28	28/170
100	270	16	44	44/270

Answers: Empirical probability approaches 44/270 or 4/25 or .16.

2. What is the mathematical probability for tossing a 3 with a dye?

List all the possible outcomes for tossing a dye. What is the ratio of tossing a 3 to the number of possibilities in the list? Answer: 1/6 or one time out of 6 tosses.

3. What ratio in number #1 did the empirical probability approach? Answer: 4 out of 25.

The above activity illustrates why the term "mathematical probability" is used. Mathematical probability is the theoretical ratio and is the ratio of S/N approaches if the "game" is fair and the number of trials is very large.

Comment: You may wish to use additional examples such as tossing coins to determine the probability for heads. (They say that the empirical probability for heads is slightly higher than the mathematical probability for heads. Try it!)

How are mathematical probabilities determined?

You have no doubt seen at a carnival a wheel with a spinner on it and the barker challenging the gullible players to try their luck. (What is a barker?) Below in the picture, what is the mathematical probability for the pointer stopping on B? G?, R?, A? (The pointer cannot stop on any of the segments, and the sections are each equal.)

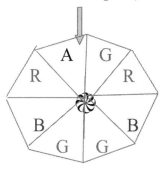

Notice the above wheel has 8 equal sections with 3 G, 2 R, 2 B, and one A. What is the mathematical probability for the wheel stopping on R? G? B? A?

Answers: P(R) = 1/4. P(G) = 3/8. P(B) = 1/4.
P(A) = 1/8

What is the sum of the probabilities? What does the sum mean in regard to the wheel? (The wheel will always stop on a wining space.)

Again, it is emphasized that in the actual spinning of the wheel the results may be very different, but if the wheel is fair and the number of cases is very large the results will approach the theoretical mathematical probabilities. Explain what a fair wheel is to your student! (How can the wheel be made unfair?)

Research: To gain an additional insight read about Pierre de Fermat (pronounced Fer-mah). (See E.T. Bell's *MEN OF MATHEMATICS*, or any book on the history of mathematics.)

There are several ways for calculating probabilities. An example of some methods is illustrated below, but first you need two postulates. (What is a postulate?)

Postulate: If the statements for successful cases are connected by OR, then the probabilities for each case are added.

Postulate: If the statements for the successful cases are connected by AND, then the probabilities for each are multiplied.

Example 1: From 3 students (we will name them A, B, C.), two are to be selected by drawing two names from a box. What is the probability of:
    Case a. That A is selected?
        Solution: A can be selected in the following ways:
           1. A and then B or
           2. A and then C or
           3. B and then A or
           4. C and then A.
           The probabilities are:
              1/3x1/2 + 1/3x1/2 + 1/3x1/2 + 1/3x1/2

$1/6 + 1/6 + 1/6 + 1/6 = 4/6$
Conclusion: P(A) is one of the names selected.) = 2/3

Case b. That AB are selected, where A is first selected and then B?
Solution: A and then B is $1/3 \times 1/2 = 1/6$

Case c. What is the probability A or C will be selected on one draw?
> P(A or C) means A is drawn or C is drawn = P(A) or P(C) =
> $1/3 + 1/2 = 5/6$, which means that 5 out of 6 times an A or a C will be drawn.

Notice "and" indicates multiply and "or" indicates add.

Chapter 8. High School Math – Grades 9-12
Section 8k. Odds, Expectation, and Fair Bets
(more Casino math)

In short the house plays the percentages, and the player relies on luck.

Gross & F. Miller
MATHEMATICS--A CHRONICLE OF HUMAN ENDEAVOR

The subject of this section is one that most students find intriguing, and even the activities are more eye opening or meaningful. In the last section, you were studied some elementary probability problems. This section will give your student some ideas as what to expect when a bet is fair or when to bet and when not to bet. The quote at the beginning of this section will have more meaning as you begin to understand terms like **fair bet**, **expectation** and the much misused term, **odds**.

It is a valid assumption that you have played a game involving probabilities at carnivals, local fairs, lotteries, received "opportunities" from the organizations where you can win $, or even at casinos with your parents. The "game" aspect is honest, but as the quote above indicates you hope to win is based on luck.

In the last section, probability is defined as the ratio of the favorable outcomes or successes to the total number of possible outcomes or trials. This was expressed in the form of a formula as $P(e) = e/N$, which reads the probability of event (e) is the ratio of the favorable outcomes for (e) to the total number of possible outcomes(N). It was stated before that the **empirical probability** (the result of actual trials) usually only approaches the **mathematical theoretical probability** after a very large number of trials. But, as you play the "game" you are playing just a few trials and therefore you are interested in what to expect from those trials. This is called **EXPECTATION**. This can be arrived at from the formula $P(e) = e/N$ by multiplying each side by n giving $nP(e) = e$. This read the number of trials times the probability for a win equals the number of favorable outcomes or the expected number of favorable outcomes.

Definition: The **EXPECTATION** for an event in a set of trials is the number of trials times the probability of the favorable outcomes for e.

$$\text{The formula is: } E(e) = n[P(e)]$$

Example: The probability of tossing a head with a coin is 1/2. The formula is P(h) = 1/2 = s/n and the expectation formula is n{P(h)} = E(s). The expectation for tossing a head can be written as n(1/2) = E(s). Now if you toss a coin 6 times E(h), the expected number of heads should be 6(1/2) or 3. But this is the expected result and may not be the actual result. Try it with a coin and check your results.

A fair bet results when in theory nobody wins. In other words, the winnings equal the cost of playing the game. This can be written as a formula also.

$$n(\$T) = n(w\$)[P(w)]$$

This reads as follows: **The cost per trial to play ($T) times the number of plays (n) equals the payout per win ($w) times the probability P(w) of winning times the number of plays (n) or the expectation.** (This really says the winnings ($) equals the money spent to play.)

Definition: A game is fair when the cost of playing equals the expected value times the winnings, or
n($T) = n($w)[P(w)], where:
$T is the cost per play.
n is the number of plays.
$w is the payout for a win.
P(w) is the probability of a win.

Example: Using the coin example above, if the coin is tossed 6 times, theoretically, 3 heads should have resulted. If the play cost $1 and each time a head turns up you win $2, then is the game fair?

$$n(\$T) = \$w)nP(w) \text{ substituting we have}$$
$$6(\$1) = (\$2)(1/2)6$$
$$\$6 = \$6$$

This indicates the game is fair, but it doesn't mean you will break even!

Before you start on the activities, we need one more definition. You often hear the term odds, like the odds of winning are 5 to 2. Most people do not use the term correctly or interpret its meaning correctly, because they do not understand the definition. Odds can be for or against an event, and are easily arrived at, if you know the probability of the event.

Example 1: If the P(e) is 5/7, the odds in favor of e occurring are 5 to 2.

Definition: The formula for odds in favor of an event e is:
Odds = P(e)/[1-P(e)].

Example 2: The probability for event e is 5/7. What are the odds? Write the formula for odds and substitute the known values.

$$\text{Odds} = P(e)/[1-P(e)].$$
$$\text{Odds} = (5/7)/(1-5/7) = (5/7)/(2/7) = 5/2$$
The odds in favor of event e are 5 to 2. (Do you see the easy way to determine the odds?)

Ask your questions, seek help, this is a good lesson for future situations and may be confusing at first.

Activity for a better Understanding

1. What are the odds for (e) given the following probabilities?
   a. P(e) = 5/7   b. P(e) = 5/6   c. P(e) = 4\7
   d. P(e) = 6/11   e. P(e) = 8/13   f. P(e) = 12/13
   Answers: a. 5/2 b. 5/1 c. 4/3 d. 6/5 e. 8/5 f. 12/1

2. What are the probabilities, given the following odds?

   a. Odds = 3 to 1   b. Odds = 3 to 2   c. Odds = 4 to 3
   d. Odds = 5 to 4   e. Odds = 7 to 3   f. Odds = 8 to 3
   Answers: a. 3/4 b. 3/5 c. 4/7 d. 5/9 e. 7/10 f. 8/11

3. If the odds against an event are 3 to 5 then what is the probability for the event?
   Answer: 5/8

4. If you toss a coin 42 times, what is the expected number of tails?
   Answer: 21

5. If you toss a dye 42 times, what is the expected number of times a three will turn up?
   Answer: 7

6. If you toss a dye 42 times, what is the expected number times a two will turn up?

Answer: 7

7. In activities 5, 6, if a toss will cost 25 cents, what should you win if the game is fair and you toss the die once?
   Answers: 5. $1.50   6. $.75

Practical Case

In a contest, the local car dealer agreed to give the lucky person a new $10,000 car, and the local bike dealer agreed to give the second lucky person a $500 mountain bike. The president of the Chamber of Commerce comes to our class and asks for your help. He informs the class they will sell only 500 tickets and only one ticket per person. The first ticket drawn wins the bike and the second ticket drawn wins the car.
   a. What should each ticket cost in order to make at least $2000?
   b. What is the probability of winning the bike, which is given on the first drawing?
   c. What was the probability of your name was drawn for the car on the second drawing knowing you did not win the bike.
   d. What is the probability of winning the car or the bike?
   e. What is the probability of not winning?
   f. Would you buy a ticket? Justify your answer.
   g. Can you ever have a probability of 2? Why?

   Answers: a. $25   b. P(b) = 1/500   c. P(c) = 1/499
         d. 999/249500 = .004004   e. .996

213

      f. No, using the expectation formula for each win, the amount is $21.04 (1/500 x 500 + 1/499 x 10000) instead of $25. But this would probably be considered charity and you may wish to help the local project.
      g. NO (see the definition)

This section has provided an opportunity for your student to create a summary. (note taking!)

Research: The life of Cardano. See the book:
    CARDANO, THE GAMBLING SCHOLAR:
                      Princeton University Press

An interesting problem

See the Jordan Curve Theorem #38 in Index 3 for an explanation.

The following puzzle is at most state Fairs in the form of a house or at a maze where you enter and try to find the exit. (In Montana ranchers will use a 2 - 3 acre corn field with paths or walk ways cut to form the maze during October as Halloween activity.) See the Jordan Theorem in Index 3 for a solution method.

Try to find your path first and then the Jordan Theorem method.

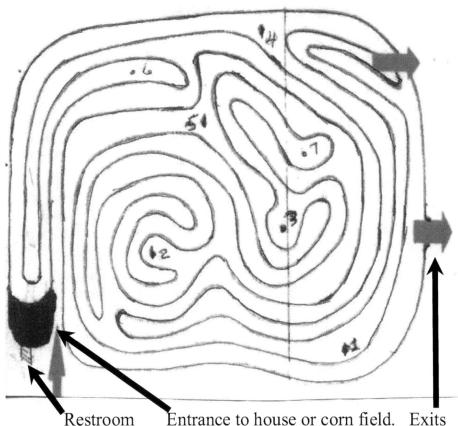

Restroom   Entrance to house or corn field.   Exits
The numbers indicate where the snack areas are.

Questions:
    a. Which snack areas are accessible? (See Jordan's curve Theorem in Index 3.)
    b. Did you try the right or left hand solution method?

Another Challenge Problem

On a hot summer afternoon, a Montana senior high school student went fishing in a nearby small pond. (A favorite Montana pastime) and while waiting for a bite on the line, he observed the following. A reed in the center of the pond,

215
when the wind blew, the tip of the reed just touched the edge of the pond. He was amazed and wonder if he could determine the depth of the pond. The figure below illustrates the problem.

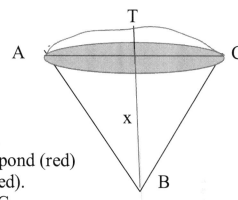

Known:

TB is x + 1 ft.
x is depth of pond (red)
AC = 10ft. (red).
TB bisects AC
ATC is an arc of a circle.

Solve for x, the depth of the pond.

Hint: Let x-1 = the depth of the pond.
(Use the Pythagorean Theorem.) Hint: $5^2 + x^2 = (x+1)^2$

Answer: 12 ft.

Chapter 9. Logical Decision Making - Grades 9-12
Section 9a. Use and Misuse of Statistics

Statistical thinking will one day be as necessary for efficient citizenship as the ability to read and write.
H. G. Wells

The education of students has not yet reached the goal stated by H. G. Wells, but this chapter is a step in the right direction.

A few Examples of the use and Misuse as to conclusions from polls.

Does your student understand what a poll is?

The following questions were asked to an unspecified random set in regard to the issue of Gun control.
  1. Would you favor a background check?
     92% at the store, 87% at gun show, 75% private seller
  2. Would you favor the following for gun safety?
       69% for gun registration
       56% for ban on assault guns
       52% ban on Ammo sells
  3. Would you favor or oppose armed guards in schools?
       54% favor   45% oppose
  4. Do you own a gun?
       49% Yes   49% no
  5. Who or what is blamed for gun violence in the USA?
       37% parents   47% social media
       23% availability of guns   60% Congress

6. Another set of results from a survey was stated in very fine prints and hard to read. The survey was listed as international, conducted by phone on Jan. 14 & 15 (Sunday and Monday) of **814** adults, all Americans, selected at random. Would you define this as an international poll with valid conclusions based on 814 adults, all American?

Questions to discuss with your student.

What would the readers like to know about these adults in the gun control poll?

    1. Were they hunters?
    2. Where they live?
    3. Age and how the age groups voted.
    4. Related Professions or careers.
    5. Education and background
    6. What is their favorite TV news show?
    7. What is their favorite radio news show?
    8. Others you may have?

Statistics is used to provide an explanation or picture of the past, partly based on the saying: That a picture is worth a thousand words. This gives us insight to part of the problem. Different viewers will interpret what they see differently. The picture below points this out? Does your student see a young woman or an old woman or both? This was created by W. Hill and appeared in Puck magazine in 1915.

Picture: Old or young woman?

Statistical methods today are influencing the decisions about the future in the medical, business, sports, agricultural, biology, economics, education, electronics, physics, psychology, sociology, chemistry, space programs, military programs, etc. H.G. Wells stated the need, but John Dewey stated a concern.

Factual science may collect statistics and make charts. But its predictions are, as has been well said, but the past history reversed.

John Dewey

Research: Who was John Dewey? What did he mean by his quote?

In this Section, you will review with your student the following which may influence decision making.

Descriptive Statistics consists of:

1. Graphs
   a. line   b. bar   c. circle   d. icon
   e. Normal curve
2. Measures of Central Tendency
   a. Mean or arithmetic average
   b. Mode   c. Median
3. Measures of Dispersion or spread
   a. Range   b. Standard Deviation
4. Polls   a. Random Samples   b. Possible Errors
5. Potential concerns leading to false interpretations of statistics:
   a. Were the questions ambiguous or misleading?
   b. Was the data random?
   c. Was the sample large enough?
   d. Were the generalizations valid and did they follow logically?

Activity to help your student's Understanding

Note: See Index 2 for definitions of Mean, Mode, Median, standard deviation.

Prepare a brief statement answering:
   When did H. G. Wells and John Dewey live?

1. It is estimated that:
   70% of the middle age population have brown hair,
   15% have blond hair,
   10 % have black hair,
   5% have red hair.

a. Show your interpretation of the above by a line, icon and/or bar graph.
  b. Interpret the above using a circle graph.
  c. Which type of graph do you feel is easier to interpret? Why?
  d. Show your graphs to 10 of your friends and poll them as to which one they find easier to understand.
  e. Report your findings to "d", to your parents, and/or teacher.

Comment: The friends may violate randomness, and also bias your conclusion. **(Help your student to understand this comment.)**

2. Two small classes have the following scores on the same test: Class 1: 8, 9, 10, 9, 9   Class 2: 7 ,8, 10, 9, 11
  a. Calculate the three averages for each class.
  b. The principal informs the parents the two classes have the same average. Is he correct and to which average is the principal referring (mean, mode, or median)?
  c. Which class do you think would be classified as better?
  d. Should the mean be used to compare small classes?
  e. What is some additional information the principal could have given to the parents in your opinion?

  Answers:
   d. Class 1: Mode 9, median 9, Mean 9.
      Class 2: Mean 9, Median 9, no mode.
   e. Probably the mean which is 9 for each class.
   f. Why do you think the class would be better?

g. No, since single numbers like the mean have less meaning and are influenced by high or low scores.
h. What is the maximum score for the test?

(How the teacher may use the test scores: Let's say the test had a maximum score of 12. How will the individual scores be used to help future students. Has this test been given before? Were the weaknesses and strengths the same? Can the test be used to improve the teaching?)

3. Calculate the mean, mode, median, standard deviation, and the range for each of the following (See Index 2, definitions 33 to 37 if needed):
    a. 1, 2, 3, 4, 5   b. 11, 12, 13, 14, 15
    c. 1, 4, 7, 10, 13

Answers: a. Mean = 3, mode = none, median = 3
standard deviation = 1.4, range = 4
b. Mean = 13, mode = none, median = 13
standard deviation = 1.4, range = 4
c. Mean = 7, mode = none, median = 7
standard deviation = 4.2, range = 12

4. If the data in number 4 represented three groups of students all taking the same course, which group would be the hardest to teach. Explain your answer.

Answer: Probably group c, since it has the larger range.

5. A student received a score of 92 on a science test with a SD of 5 and a mean of 87. The same student scored an 82 on a math test with a SD of 3 and a mean of 75.
   a. Which class did the student do better in with respect to the other students? Explain your answer.
   b. Which class has a wider range of ability? Explain your answer.
   Answers: a. Math test   b. Science class

6. The following is from a local paper during 1997 in Montana: **Young Americans are confident, optimistic about their future.** A survey of 334 youngsters ages 14 to 18 by a pollster states that teens are optimistic and determined as they enter a new school year.

   ### Another poll

   A national telephone survey of 2,001 teenagers states: **The teenagers want good pay, continue their education, job security, to be creative, and avoid routine work.** This article then used 12 inches of type explaining what the youth hope to do in the future and what they didn't want to do. Are the responses what you're your student would expect?

   What are some of the questions the article should have answered to make the surveys more creditable and informative?

   Some questions: Location of study? How students were selected? What time of year? What type of questions and how worded? etc.

7. A student was tossing a coin and in 10 tosses a Head never turned up. The student said: "By the law of averages a Head should turn up on the next throw." Then the student tossed the die. Another Tail turned up!
   a. Write your thoughts as to the student's comment.
   b. What is wrong with his comment?

   Answer: The coin has no memory and consequently the probability of tossing a Head is the same as it was for all previous tosses.

8. Write the next row of numbers in the following and look for a pattern.

   ```
                     1
                  1     1
               1     2     1
            1     3     3     1
         1     4     6     4     1
      1     5    10    10     5     1
   ?     ?     ?     ?     ?     ?     ?
   ```

9. Expand the following. And write the coefficients;
   $(A + B)^0 = 1$
   $(A + B)^1 = A + B$ 1   1
   $(A + B)^2 = ?$ _   _   _
   $(A + B)^3 = ?$ _   _   _   _

   a. Look at the coefficients and compare them with the results with part "a."

Comment: This is called Pascal's Triangle and is a clever way to determine the coefficients for binomial multiplication.

b. Does this apply to probabilities? Yes, it does. Take the case of tossing coins. If there is one toss of one coin, the probability for ahead is ½ and for a tail is ½.

H or T. (1 1)

If there are two tosses of one coin, the possibilities are:

TT or (TF or FT) or FF
1   2   1

If there are three tosses of one coin, the possibilities are:

TTT   (TTF TFT FTT) (TFF FTF FFT)   FFF
 1   3   3   1

c. How do probabilities use Pascal's triangle? For one toss the probability for T is ½ and for H is 1/2.

For two tosses the resulting in TT, 2TH, HH, and the probability for each is: P(2H) is 1/4, P(TH or HT) is 1/2, and for two Tails P is 1/4.

d. Using the same method show the case for three tosses.

e. Would the answers to "d" be the same for two coins tossed at the same time? Three coins tossed all at the same time? (P is probability)

Answers: P(3H) = 1/8, P(2H1T) = 3/8, P(2T1H) = 3/8, P(3T) = 1/8.

This is certainly an easier way then you used in the section on probability.

INVESTIGATION
Gauss' problem

In the previous sections your student using the process of inductive reasoning discovered formulas for the sum of N odd integers and also one for the sum of N even integers. The famous mathematician K. Gauss was in an elementary school class in Germany (early 1800) and the teacher gave the problem to the class to find the sum of the first 100 counting numbers (1+2+3+4+5 and on to 100). What is your method and answer? Hint: The following table is what I think Gauss did and he observed a PATTERN.

CASE	THE SET	THE SUM
1	1	1
2	1+2	3
3	1+2+3	6

continue the table until you see the pattern.

Answer: Sum = $N(N+1)/2$ or 5050 for the first 100.

Chapter 9. Logical Decision Making - Grades 9-12
Section 9b. Inductive Conclusions
(A MUST!)

Rodin's The Thinker,
Palace of the Legion of Honor, Lincoln Park, San Francisco

Your student may recall what inductive reasoning is, and the weaknesses involved in this type of conclusions. This chapter will provide a review of these Decision Making methods and add a few more.

Someone once stated the following, probably a Math teacher:

We learn the new in the light of the old.

Anonymous

What is Inductive Reasoning?

In a previous chapter your student did some inductive reasoning activities and may recall the weaknesses. Your student may recall, inductive reasoning was defined as a method of reaching a general conclusion by observing or repeating, a few cases, and then assuming the rest of the future cases followed the same pattern. Another way to simply state the concept, would be that a conclusion is reached from observing a few cases. In your geometry course, you probably drew a few geometric figures, then stated a conjecture and with the help of your teacher or mentor you tried to prove the conjecture. Once the conjecture was proven, you classified it as a theorem. Remember?

This section will be using inductive reasoning to make conjectures and emphasize that they are based on assumptions.

A few cases are listed below for you to practice making conjectures.

Understanding Activity

Your student's conjectures may come up with different answers than are expected. This makes for an interesting activity. Ask your student to explain their conjectures.
1. 1, 1, 2, 3, 5, 8, ?, ? What are possibilities for the (?s)?
2. Subject M9 was boring, therefore subject M10 will be boring.
3. My school always beats your school in football, therefore we will win this year.
4. **(A must!)** After observing a few simple cases, write a conjecture for the sum of the first 10 even counting numbers and then the sum of the first N even counting numbers.

Hint: 2 + 4 = ?, 2 + 4 + 6 = ?, 2 + 4 + 6 + 8 = ?

Cases	N	the Sum is?
1	2	2
2	2 + 4	6 continue until you see the pattern for n cases.
3	2+4+6	12 (N)(n+1) = sum

5. Write a conjecture for squaring any number ending in five (5). Write a few cases and look for the pattern.

Number	The square
5	25
15	225
25	625 (2x3) 25
35	? Ans: (3x4) 25
n5	?

6. The "divisible by three" conjecture:
   a. Write ten two-digit numbers (ex. 12, 30, 99) that are divisible by three.
   b. Add the digits for each number in part a. If the first addition is a two-digit number, add again. Example: 78 --> 15 --> 6
   c. Write a conjecture.

7. The "divisible by nine" conjecture is:
   a. Write ten 2 5-digit numbers which are divisible by nine. The last two should be 5-digit numbers. (Use your calculator.)
   b. Add the digits for each number in part a. If the first addition is a 2-digit number, add again. Example: 1368 --> 18 --> 9
   c. Write a conjecture.

Answers to problems:
1. The next two are 13, 21.

2. 2. False conclusion. A previous feeling about a course has no bearing (or very little) on the next course.
3. S = 2n-1
4. Pass performance may have little to do with the future.
5. N(n+1) (25) e.g. 35 squared is (3x4) and 25 or 12 25
6. If the digits add to a number divisible by three, then the original number is divisible by three.
7. The "divisible by nine" conjecture
    a. Write three 6-digit numbers which are divisible by nine.
    b. Add the digits for each number in part a. If the first addition is a two-digit number, add again.
    c. Write a conjecture.

Comment: Review of a previous method

This is called **Casting out Nines.** This method was actually taught in the schools before calculators were available as a means of checking answers. This method works for all four basic operations. Suggestion: Explain the method to your friends by working examples for them.

Example: Method for checking addition problems.

796 adds to 22 adds to 4
835 adds to 16 adds to 7

Sum is 1631 adds to 11 and the 7 + 4 above adds to 11, therefore, the answer is correct.

Example: Method for checking subtraction problems.

$$\begin{array}{l} 1636 \text{ adds to } 16 \text{ adds to } 7 \\ \underline{- 796} \text{ adds to } 22 \text{ adds to } 4 \;\; (7\text{-}4 = 3) \\ 840 \text{ adds to } 12 \text{ adds to } 3 \end{array}$$

Therefore, the answer is correct.

8. Euler's Polygon-Region theorem
   a. A triangle has how many **segments, vertices** and **regions**?

   V = 3
   S = 3
   R = 2 (inside and outside)

   Pattern is V + R = S + 2
   V = 4
   S = 5
   R = 3

   V = ?
   S = ?
   R = ?
   Is V + R = S + 2 ?

   Does the same pattern work for 3D figures?

   b. Draw two or three 3-D figures and observe if the pattern is valid.

Interesting research project: The life of L. Euler

Chapter 9. LOGICAL DECISION MAKING, Grades 9-12
Section 9c. Illusions, Games and Applications

Rodin's *The Thinker*
The Legion of Honor, San Francisco, CA

The important thing is (to) not stop questioning...
Albert Einstein

In this section, your student will review inductive reasoning and some of its weaknesses in **non-mathematical** situations that your student will encounter. The weakness in inductive reasoning is in the conclusion assumption or conjecture, which is a statement about the future made from a few cases. Superstitions are consequences resulting from inductive reasoning usually from observing a few cases. We naturally draw conclusions from the cases, but we must realize the weakness of predicting from these few cases. Many times the conclusions are from visual cases and what we think we see is not always the real case. Missouri is known as the

232

"show me" state, but a person can see things differently than another person.

Remember this type of reasoning (inductive) is probably used more than any other type. It could be thought of as "jumping to a conclusion." The following examples were collected from NCTM conventions.

Examples:
    1. Did you ever see the arrow (--->) on the Fed EX truck? Once you see you will never forget it. In fact, you will see it every time you see one of their trucks. The following will help.

Hint: Look between the second e and the X.

    2. Is the top hat taller than it is wide? Guess first and than measure it.

Is AC > BD?

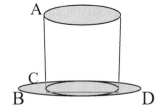

    3. Do you see 3 or 5 boxes? Most students see only 3, but look at it when the picture is inverted.

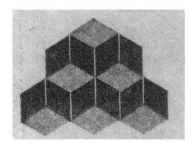

4. Refer to the next figure and where is the small box, In the corner or outside the big box? Blink a few times while looking at the figure and you may see the two set of boxes.

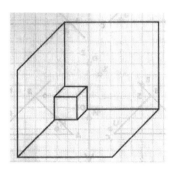

5. Have you stood on railroad tracks and looked down the tracks? What do the tracks appear to do?

Answers: 1. AB appears longer than CD, but they are equal.
2. The hat is wider.
3. Keep looking until you see the 3 cubes and the other case 5 cubes.
4. The figure appears to have the small box in the corner, but by moving your head to the right or left of the box you may suddenly the small box outside the large box.
5. The tracks seem to merge at a distance point.

Game Activity

1. The game of Sprouts* for two players. (parent and student)

*This game was invented by J. H. Conway and M. S. Paterson in 1967and will take some time for the parent and student to learn to play. (A teacher could use a class period to teach the class, divided into teams of 2, how to play it.)

a. The **objectives** is to be able to predict: (1) who will win the game, the player who makes the first move or the player who makes the second move.
b. The second objective is to be able to predict the number of moves to complete the game given the number of seeds to begin with.

Rules:
1. Draw a curve (call a sprout) from a legitimate seed to another **legitimate** or same seed and then place a **new seed at the midpoint** of the sprout, which will then become a new seed.

2. A **legitimate** seed is any seed which has only one or two sprouts or zero sprouts on it.

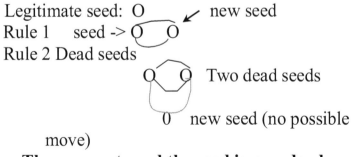

Legitimate seed: O   new seed
Rule 1   seed -> O O
Rule 2 Dead seeds
                    O O   Two dead seeds
                    0   new seed (no possible move)
Three sprouts and the seed is now dead.

3. The **winner** of the game is the player who makes the last possible move.

Start the game by placing a point (seed) on the paper and complete the following table after the moves.

Example: Start with 1 seed

Starting # of seeds	Players		Number of moves		Winner (A or B)
	A	B	A	B	
1	O		1		

The picture of the first move by A is: New seed (#2) at midpoint of arc.

The above shows that player A made the first move and created a new seed on the arc, O.

Now B's turn: B's new sprout and new seed

Now A's turn but there is no possible move, so B wins! (Do you see why?) The record is below.
It may take some practice to learn how to play.

Your record for the 2-seed game should be as below after you completed the game. (Complete the table for starting with 3 and 4 points. The following is one set of possible moves.)

Record the Game	Moves		Total moves	Winner
	A	B		
Start with 1 seed:	1	1	2	B
Now start with 2 seeds:	1	?	?	?

If the game starts with 1 seed, then there will be 2 moves and B wins?
   a. If the game starts with 2 seeds, then there will be? moves and who wins?
   b. If the game starts with 3 seeds, then there will be? moves and who wins?
   c. If the game starts with 4 seeds, then there will be? moves and who wins?
   d. If the game starts with 5 seeds, then predict the? number of moves and who wins?
   e. If the game starts with an even number of seeds, then predict the number of moves and who wins?
   f. If the game starts with odd number of seeds, then explain who wins and the number of moves.

Comment: The game can be played with three persons, but the generalizations will be different.

Suggestion: Play this game with your friends and they will be mystified, since you will know who will win and the number of moves it takes to win.

Another interesting application

1. The shortest distance between 2 points is a straight line? In Taxicab geometry, the statement is not true.
   How many possible ways does the cab driver have to go from A to B in case a? Rule: The cab must always go in the direction of B.

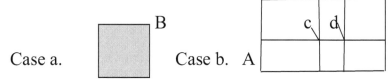

Case a.   Case b. A

Case a. Ways to go from A to B? There are 2 ways.
Case b: Ways to go from c to d? Your answer is? (1)
Case b: Ways to go to c from B? Answer is? (3)
Case b: Ways to go to B from A? Answer is? (10)

2. Can you draw a curve using only straight lines? Most people say no. In the following figure connect 1 to 8, 2 to 7, 3 to 6 and continue until for all 8 connects.

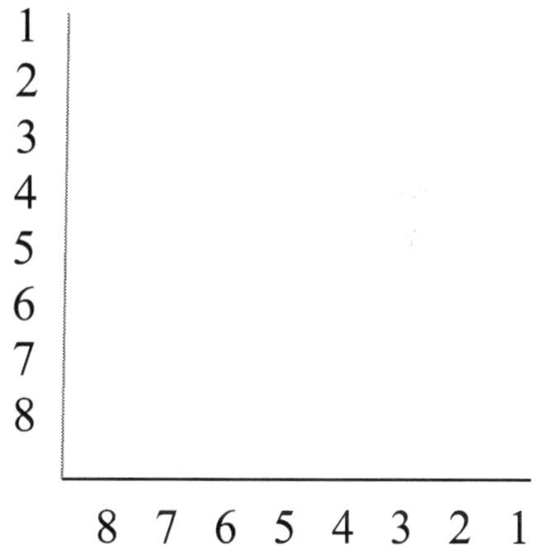

```
1
2
3
4
5
6
7
8
        8 7 6 5 4 3 2 1
```

The Cycloid

This is an interesting figure. It is also the curve of quickest dissent. (Parent and their student will like this and should work with each other.)

3. Two wheels (circles) of different size (radii) are bolted together so they are concentric. Corresponding points A and A_1 are labeled as in the figure below. Now if the large wheel is rotated one revolution along the line, the distance is the circumference and the small wheel will also make one revolution and hence you could conclude

the circumferences of both circles are the equal. But how can two different size circles have the same circumference? **Note: A and A1 are points on the circumference of the 2 circles.**

4.

a. Make a model of the above for demonstration and illustrate what actually happens when the wheel is rolled* with regard to the paths for the two points.

Hint: Make a hole at A large enough to insert the tip of a pencil and trace the curve for point A as the large circle completes one rotation.

b. Using the same procedure as in "a" but for point A_1, trace the curve.
c. Observe the two curves and you will see the solution!

Answer: The curve generated by A is a cycloid. The physical property of the cycloid is that it is the curve of quickest descent.

Research: Use the computer and the internet and do search for information pertaining to cycloid.

Record your notes

Chapter 9. LOGICAL DECISION MAKING, Grades 9-12
Section 9d. Forms of an Implication and their uses

The advertising world and other "worlds" (such as politics) use these forms of an implication to take advantage of people who do not understand implications and their possible misinterpretations.

The following diagram will be used to help you understand the forms and their meanings.

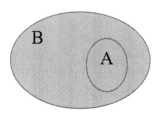

This diagram is to be read as: If A, then B or If you are in A, then you are in B.

Look at the diagram and answer the following statements.
    Original statement: If A, then B (consider this valid and true).
    Converse: If B, then A (not always true or valid).
    Inverse: If not A, then not B (not always true or valid).
    Contrapositive: If not B, then not A (true and valid).

Implications are very important! Be sure your student understands how they are used and their interpretation possibilities from the following!

    a. Fact: If you fly the Flag (A) on Memorial Day (B), then you are patriotic. Assume the fact is valid, then which of the following are valid? To help you make a decision, translate each into if A, then B form.
        a. If you are patriotic, then you fly the flag on Memorial Day. Name? (Converse)
        b. If you do not fly the flag on Memorial Day, then you are not patriotic. Name? (Inverse)
        c. If you are not patriotic, then you will not fly the flag on Memorial Day. Name? (Contrapositive)

        Answers: a. Converse, may not be valid.
        b. Inverse, may not be valid.
        c. The Contrapositive is valid and true.

    b. Fact: If you watch TV-PBS, then you want factual information. Assume valid, then which of the following **may be** valid?
        Converse   Inverse   Contrapositive

    c. Fact: If you want factual information, then you watch TV-PBS. Which of these are valid? Name each.
        a. If you don't watch TV-PBS, then you don't want factual information.
        b. If you don't want factual information, then you don't watch TV- PBS.

c. If you watch TV-PBS, then you want factual information.

Answers. a. Contrapositive b. Inverse c. Converse (Only "a" is valid, but b and c may be valid.)

Data Collection

Listed are questions you should know about polls. Give an example of each of the following and how it could influence the result.

a. The number polled is important. (Many times you are never told the details. The number polled is important and is it truly random! Where and how collected is very important. If you want a particular outcome from your poll you select a certain sample.

Example: If you want a positive output relative to guns you may select NRA members for your random sample. (Which would not be random.)

b. The selection process, time of day, the wording of the question, randomness, age, location are very important.

Example: If you want a GOP politically inclined result then you select a known GOP community or city.

c. The method of contact (email, telephone, U.S. mail, or interview.) can have an impact on the outcome.

242

Each of these could limit randomness. Good for a discussion with your student!

Two witnesses can report completely different information as to what they thought they saw or heard. (The following was sent to the author by a friend.) Witness A reported seeing a young woman. Witness B reported seeing an older woman. Two polls or witnesses can report completely two different conclusions. (Look from the left and then from the right. Puck Magazine 1915 by W. Hill)

In the figure below, which segment is longer AB or BC?

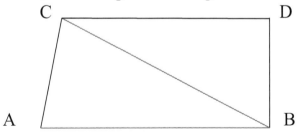

Measure each - they are both the same length, but not in appearance.

A Classic Random Sample from History

The 1936 Presidential election is a classic example. The election was between F. D. Roosevelt and A. Landon. A large random sample predicted that Landon would easily win by a landslide. It turned out just the opposite. The question was why, or what went wrong?

It turned out the random sample was from the set of voters who had telephones. Most voters did not have telephones during the 30's at that year, but they knew where the voting booths were. The sample was random, but not from the correct set of people.

Notes

Chapter 9. LOGICAL DECISION MAKING - Grades 9-12
Section 9e. Basics for Decision Making

The key to understanding and making logical decisions is that all conclusions are based on four items:
Undefined terms
Defined terms
Assumptions
Conclusions based on the above three, valid or not!

Example: 30-50% of the words we use or write are undefinable. In the following sentence { }, there are 4 undefined terms. They are highlighted. (Four of the 8 words may be classified as undefined or 50%.)

{The Declaration of Independence is a perfect example.} In the above statement there are 4 (red) undefined terms.

Suggestion: Read the first few pages **with your student** and observe why independence from England was justified.

Remember statements are true or false, but conclusions are valid or invalid.

About 45 % of the words in any statement are undefined!

The following are used in decision making, some are valid methods and some rigged to obtain a desired result, but all require an informed decision maker to really evaluate them. All parents and teachers want their students to be valid decision makers!

> **Guessing**
> **Listening to experts or non-experts on all sides of issues**
> **Observation**
> **Induction**
> **Estimating**
> **Direct reasoning**
> **Indirect reasoning**
> **Statistics**
> **Data collection**
> **Forms of an Implication (Statements, Converses, Inverses, Contrapositives)**
> **Advertising**

(There was a tobacco ad at one time that suggested the following: Try our product for 30 days and then if you don't like smoking quit). What was wrong with this ad? In 30 days, you are addicted.

Discuss with your student the above using examples from ads or your past experiences. In the following, ask your student to list the words which need defining?

Preamble To The Constitution

We the people of the United States, in order to form a more Perfect Union, establish Justice, insure domestic Tranquility, provide for the common defense, promote the general Welfare, and secure the Blessings of Liberty, to ourselves and our Posterity, do ordain and establish this CONSTITUTION for the United States of America.

Are there any words in the Preamble that you would classify as undefined terms? Count them and list them list the words which need defining in order for really understanding the meaning. Your list will naturally be different than your student's.

Do the same with the Pledge to the Flag.

Notes and comments

Chapter 10. The Student's Future
Comments for the Parents and Teacher

Rodin's THE THINKER at Golden Gate National Recreation Area in San Francisco, CA

The author has attempted to assist Parents and Teachers by the method "Learn the new in the light of the old" and with some math History, plus to answer many of the questions your students may have. One of these questions students always have is: "Isn't there an easier way to work the problem?" It is a natural question, and the history of many disciplines verify this. The author tried to show this in various ways. A key to learning is, reading the material (perhaps several times), taking notes, seeking help, understand the errors, practice and more practice, and reading some of the past history of a topic, plus let us not forget Parent -Teacher- Student team work. The history makes the topic more interesting. Let us not forget that encouragement and understanding is a must by the parents and teachers for **student success.**

We all want our students to be understanding Thinkers and Valid Decision Makers. This is why the author incorporated into this book what all decisions are based on with many everyday examples, plus examples of the misuses and how many errors are made. Examples can be found in ads, news reports, TV shows, magazines, and newspapers every day.

Your student will find the course very informative, useful and to varying degrees, even enjoyable at times. One major factor is that it will be beneficial for any area, graduation, college, tech school, trade school, and employment. Let us not forget the help it will provide for the SAT or ACT higher scores and future decision making for life situations.

Now days the education doesn't end when the student graduates from high school or college, it is a continuing life time requirement.

Remember the bibliography of selected titles provided and suggestions for additional readings that are suggested throughout the text. You and your student will fine many very informative and enjoyable books listed in the bibliography. The author knows what your first thoughts were, but suggests you try a few more of the books and your thoughts will change!

The Objective

This book is for the benefit of students in grades 1-12. This benefit will be maximized if their teachers, and especially the parents, assist the students in what is called the Team Work Approach. This book is written for parents and teachers with the objective for better students and citizens.

Why the grades 1 to 12? These are key years where many students, due to their expanding world, will be left behind if they are not encouraged and helped with their learning, especially grades 6, 7, and 8. This book will aid the teachers and parents by providing bits of history as to the development of mathematics, resulting in learning and understanding a bit of the Wonderful World of Mathematics. It will assist the parents to understand what is being taught and consequently help the student understand. It has been stated in many ways that you can tell the extent of nation's civilization by the extent of mathematics obtained by all citizens. As a civilization becomes more sophisticated, especially in a Democracy, the people must become more educated in the area of decision making. Today's demands require a life time of learning, which is always changing and more demanding. Look at the number of new applications we have been challenged to learn and operate in the last 40-50 years. (This makes for an informative discussion, ask your students to list a few, even to talk with the Grandparents.) Schools have made a mistake to not offer special programs to keep parents informed, especially in the early years as to the changing methods and content, so the parents can help their children succeed. The parents become almost a teaching assistant, and may need some background in math. The concepts are introduced in a historical, meaningful, explanatory way that will help the parents and teachers to be more helpful for their very important objective, plus make the learning more interesting. The author realizes the parents may feel frightened by this approach and the author would be wrong if he stated it would not require some extra work. An interesting addition in this book is the Bibliography of source books for teachers, parents and students to make the course more interesting plus providing help for both the teacher, parents and the students. You will find these books very interesting if you give them a try. You will be proud of the end result with your student. Teachers and parents will be surprised at the enjoyment in reading some of these books. (You probably don't believe it, but try it! Suggestion, try a few, like L.C. Hogben's books.)

Chapter 11. Course Review – Grades 10-12
Section 11a. Geometry

The proof that understanding Geometry has been the great practice field for logical thinking… and has been validated that its provides the basic training for the student in Decision Making training (if taught correctly).

Harold Fawcett
Nature of Proof, 13th Yearbook of NCTM

Geometry Review

The Geometry content will be reviewed first since the major tests like the SAT or ACT and other college or junior college results have indicated it (Geometry) is the most needed. The following procedure or steps will help you solve the problems. (Socrates and Plato, ca. 500 B.C.E, establishes Geometry as the simplest and best way to teach decision making basics in their Academies for future leaders in Athens, the first democracy. What happened to Socrates?)

These reviews should be printed out for the student to take. Give the student all the time needed to complete. Understanding is important not speed.

Parent and Teacher Suggestion: Let the students justify their answers to you and discuss any questions that may have.

To the student: Don't forget to use the five indexes for quick reference to assist you in the solutions. Be able to justify your answers. Take notes of the problems you have trouble with.

Suggestions for the student:
        a. Draw the figure.
        b. Label the figure.
        c. Identify the unknown.
        d. Write the equations.
        e. Solve for the unknown.
        f. Answer the questions.
        g. Use your Calculator
        h. Organize your work so it is easy to follow your logic to the answer.

Questions (Use the indexes if needed)

1. Are the following statements valid definitions? Defend your answer. (A definition is valid, if the definition is true when reversed.)
    a. A restaurant is a place that serves food.
    b. Mathematics is a useful course.
    c. A postulate is an assumption.
    d. A plane triangle is a set of three non-collinear points and the line segments determined by the three points.
2. Name three geometric terms in 1d that are classified as undefined?
3. How many points are needed to determine a geometric line.
4. The points on a geometric line correspond to the numbers on the _____ line.
5. Draw a ray and label it AB.

6. Draw a line segment and label it AC.
7. Draw a triangle and label it CDE.
8. A geometric plane is determined by ___ ___ ___ points.
9. What is the sum of the angles in a plane triangle? (Theorem 3)
10. How is the distance between points A and B determined?
11. What is a theorem?
12. What are the conditions for 2 triangles to be similar? (Theorem 7)
    a.   b.   c.
13. When are triangles congruent?
14.   a. What are parallel lines?
      b. What are skew lines? (Need a dictionary?)
15. Draw three acute scalene triangles, label each ABC. (Use your ruler and protractor.) (Theorems 20,21,22)
    a. In one triangle draw the three medians.
    b. In the second triangle draw the three altitudes.
    c. In the third triangle draw the three angle bisectors.
    d. Write three conclusions in if-then form.
    e. The shortest distance from a point to a line is the ___ distance.
16. If A implies B and A is given, then ___.
17. Draw a rhombus that is not a square. (Need a dictionary?)
18. Draw: a. convex polygon. b. concave polygon (Need a dictionary?)
19. If the sides of a triangle are 46, 23 and x, then what do you know about the measure of x? (Theorem 1)
20. The number of square units a plane figure contains is its ___.

21. The Pythagorean Right Triangle Theorem states _____. (Theorem 18)
22. What is the angle sum of each of the following polygons? (No protractor)

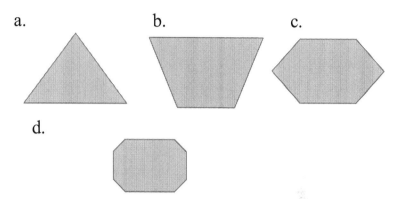

a.

b.

c.

d.

23. If each letter represents a unique digit, then what are the possible digits for the following addition problem?

      GO   Hints: G must > than ___
    +GO   O can't be ___?
    WIN   What digit is W?

24. How many geometric planes may four points determine?
25. The two top teams in the conference ended the season with records as shown in the following table.

Team	Wins	Losses
H	29	7
V	32	8

Team H defeated W once and Team W defeated H once. Which team would you say should receive the Cup. If there is no play off. Why?

26. What is the perimeter of right triangle ABC, where C is the right angle? Given: AB = 3x units and BC = 4x units (Theorem 18)
27. What is wrong with this argument?
    A yard is 36 inches.
    ¼ yard is 9 inches. (Take the square root.)
    Therefore, ½ yard is 3 inches.
28. Students of school X voted that all students will wear RED cap at the Saturday game. Which of the following are valid or not valid and why? Hint: Write the given in if-then form.
    a. John wore a red cap, therefore he is a student of X.
    b. John is not a student of school X, therefore he will not wear a red cap to the game.
    c. John did not wear a red cap to the game, therefore he is not a student of school X.
29. Draw the 3D view for a.
    a. Given: Top view   Front view   Side view

    b. Draw the layout of this figure.

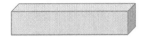

30. A football field is 50yds by 100yds and is really a large rectangle. What is the length of the diagonal?

31. Draw the **top, front and side** views of the following figure.

32. Circle A has 3 times the area of circle B. What is the radius of circle B, if A's radius is 3?
33. A triangle is formed by connecting the midpoints of the sides of an equilateral triangle, then what is the perimeter and area of the inner triangle? The original triangle has a side of 6 inches.
34. A park is in the form of a square (call it A). Another square (play area) is formed by joining the midpoints of the sides of A. The inner square will be the play area. What is the area of the play area if the side of square A is 200 feet? Draw a picture of the park. Where do you think the picnic areas are?
35. You school is erecting a new 25 ft. flag pole. What would you advise the principal as to the way to tell if it is perpendicular from all directions.
36. What is a theorem?

Answers

1. a. True, but not a definition. b. True, but not a definition. c. Valid d. Valid.
2. point, line, plane
3. Real Number line
4. Real number line
5. $\overline{A \overset{\rightarrow}{} B}$
6. $\overline{A\ C}$
7.
```
        C
   D ▲▲▲ E
```

8. Three non-collinear points
9. 180 degrees
10. The measure of the line segment.
11. A theorem is an important mathematical statement that can be proved.
12. a. Angles are equal (AA).
    b. The corresponding sides are proportional (SSS).
    c. Sides proportional and included angles equal
13. The figures are identical, angles are equal in measure and the ratio of corresponding sides is 1.
14. a. Parallel lines are lines in the same plane and do not intersect.
    b. Skew lines are lines not in the same plane and do not intersect.
15. Each intersect in a point
16. Perpendicular distance
17. Then B
18.
19.
20. $46 < x < 69$ units
21. Area
22. In a right triangle with sides a, b, and c, then $a^2 + b^2 = c^2$ where c is the hypotenuse.
23. a. 180 degrees   b. 360 degrees   c. 720 degrees   d. 1080 degrees. The formula is $[S = (n-2)180]$
24. G >4, O is not 0, and W is 1.
24. Four Planes: ABC, ABD, ACD, BCD
25. Team H has a higher winning percentage, but there could be other factors to be considered like Team W's and H's

schedules, or the scores when they played each other. Coaches could also vote.

26. 12 units
27. Listen to their explanation (not an equation).
28. c" is valid, the contrapositive
29. a. Round Cone   b.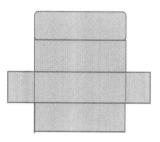

30. 111.8 yards or 111yds and 29 inches
31. Actually use a cylinder (can of soup) and look at it from each position.

        Top   Front   Side

32. Hint: AREA = 3 area! $r = \sqrt{3} = 1.73$
33. ¼
34. total park area is 40000 and play area is 20000 Sq. feet.
35. Perpendicular is two directions, then perpendicular from all directions.
36. An important mathematical generalization that has been proved.

Suggestion: If your student missed any of these problems, record the missed problem numbers, and help your student to understand their errors. Check the Table of Contents and review the material if needed. Then wait a week or two and the retake the review. This review covers the basic material. When your student completes them their understanding and satisfaction will amaze you.

The proof of the pudding is in the eating.

There is nothing wrong with missing a problem once or twice, but we should learn from the errors.

**Challenge Problem
For Parent and Student**

Where on the earth can you go 10 miles south, 10 miles east, and 10 miles north and be back where you started?
Hint: There are many places!

Comment: The first response is the North Pole, which is one correct answer. Can you find another answer? There are actually an infinite number of points in theory, where this can be accomplished. One solution: Pick a point 10 miles north of the 10-mile circle of latitude near the South Pole as another source of infinitude of points.

**CHAPTER 11: Course Review, Grades 10-12
Section 11b. Algebra Review**

Comment: Your student may find this review challenging, but isn't that they may they learn! Let the students take all the time they need. Advise them that they may even ask for your help. They may use their notes, calculator, even refer to the chapters, but should write complete answers and/or questions and review the concepts involved. (Use your calculator when needed.) Ask for explanations in class, because if one student has a question, then other students will also. Print out the review and look at it again, especially a few days before exams or your next math class, or taking the SAT or ACT. Content is considered the basic mathematical needs for your future vocation or profession, and for an informed citizenship.

Teacher: If you use this review it could take several days with discussions as to how and why. Reviews are for learning and not for grading, but they are very informative as to improvement for teaching future classes!

Suggestion: Print out the questions in part I for the student.

Part 1: Short and quick answers (Use the Indexes if needed)

Questions 1 – 7 involve number sense and skills

1. If x is 4 and y is 3 then what is the value of:
    a. $2x-3y + 9x/y - 5(x-4y) + (x-y)^2$?
    b. $x/y + y/x$?
    c. $(x/y)/(y/x)$?
    d. The absolute value of $-x$ times y?
2. What is the answer to 0/5?
3. What is the answer to 5/0?
4. If x is 1/2 and y is 2/3, then what is the value of:
    a. $x + y$? b. $y - x$?   c. xy?   d. x/y?

5. What is 10% of $17.50?
6. Give an example for the following:
    a. Addition of integers is commutative.
    b. Write a second-degree equation in one variable.
7. Short answer questions:
    a. The additive inverse of –5 is?
    b. The multiplicative inverse of 1/3?
    c. The square root of 49 is? The value of $\sqrt{49}$ is ?
    d. How far is the point (5,12) from the point (0,0)?
    e. The tangent of 45 degrees is?
    f. The sin of 30 degrees is?
    g. If the probability of A is 3/5, then what are the odds for A?
8. Write an equation that meets the following conditions.
    a. The X value is 6 for this equation, $3X + 5 = K$.
    b. What value is K?
9. In the $3X + 5 = Y$ and if you graph it, where does the line cross the y - axis? cross the x - axis?
10. What is the equation for the line determined by points (-2,3) and (4,5)?
11. What is the acute angle the line in #10 makes with the x-axis? (Nearest degree)

Part 1. Answers

1. a. 52   b. 25/12   c. 16/9   d. 12
2. 0
3. None, empty set.
4. a. 7/6   b. 1/6   c. 1/3   d. ¾
5. $1.75
6. a. 3   b. $x^2 + x + 1 = 0$ or similar type.
7. a. 5   b. 3   c. ±7, 7   d. 13   e. 1   f. ½   g. 3/2

8. 3x + 15 = 23
9. x intercept is -5/3 and Y intercept is 5.
10. y = (1/3)x + 11/3
11. 72 degrees

Suggestion: Print out Part 2 for the student.
   (Use the Indexes if needed)

Part 2. Problem Solving

1. Pete and Sally were good friends and Pete noticed he weighed 60% more than Sally. When they weighed at the fair their weights totaled 210 lbs. What is the weight of each? (Nearest pound)
2. How long will take $500 to double if invested at 12% per year?
3. Tony mows lawns to earn book money. Last Monday he mowed a yard that measured 110 feet by 200 feet. His charge is $.02 per square yard with clean up. What amount did he earn?

(This was made by a creative student! See the next page)

Source unknown

4. Student walking to school took the following short cut AC as shown below.

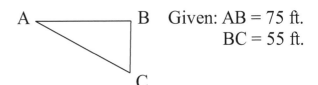

Given: AB = 75 ft.
BC = 55 ft.

   a. What is the length of AC?
   b. What is the distance saved by using the short cut AC (in feet to nearest integer)?
5. In the triangle in #4 what is angle A equal to (the nearest degree)?
6. Graph (shade the combine region) the following on one set of axes and name the shaded figure.
   Given: $Y < X$, $X < 4$, $Y > 0$
7. As the temperature increases, the cost of heating the office goes down. What type of variation is this? Direct, Indirect or neither.
8. a. How many answers does a first degree equation have?
   b. A second degree equation has how many answers?
8. Solve this equation. $X^2 - 36 = 0$
9. Write an equation whose answers are 2 and -3
10. What is the maximum point for the equation $Y = -X^2$?
11. What is the equation for the line determined by these two points? (1,2) and (-3,4)
12. If log N = 2, what is the value of N?
13. The radius of a circle varies directly or indirectly with the circumference of the circle. Which is it?
14. If you double the radius of a circle, then what happens to the area of the circle?

15. What happens to the volume of a circular cylinder if you double the radius?
16. In a 30-60 degree right triangle ABC (C is the right angle), if the hypotenuse is 10 units, then what is the length of the altitude from C?
17. If a baseball has a radius of 1.5 inches, then:
    a. What is its volume? (Nearest cu. in.)
    b. What is its surface area? (nearest sq. in.)
19. A parallelogram has sides of 10 and 8 and one of the angle is 60 degrees. What is the area (to the nearest inch?
20. In a right triangle, the 2 acute angles have the same value for the sin of each acute angle? What do you know about the name of this Triangle?

Part 2. Answers

1. 129 = P, S= 81 lbs
2. 6 years
3. $49
4. a. 93 ft.   b. 37 ft
5. 36 degrees
6. See graph ⇨

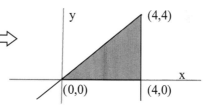

7. Indirect
8. a. 1   b. 2
9. 6 and -6
10. $x^2 + x - 6 = 0$
11. (0,0)
12. y = -x/2 +5/2

13. n = 100
14. Directly
15. Area is 4 times larger
16. Volume is 4 times larger.
17. $(5\sqrt{3})/2$ or 43
18. a. 14 cu. in. b. 28 sq. in.
19. $40\sqrt{3}$ or 69 sq. inches
20. Isosceles right triangle.

Note: Make a list of problems your student missed and review them and rework them in a few days.

A few interesting problems
(Guess first, then solve using the hints)

1. Justify your answers to the following (hint: See Index 3 for needed formulas):
    a. a. Can the area of a circle ever be the same number as the circumference of the circle? Hint: $2\pi r = \pi r^2$ and simplify.
    b. Can the area of a square ever be the same number as the perimeter of the square? Hint: $4S = S^2$

    Answers: a. Yes, when r is 2. r   b. Yes, when s = 4

2. Can two concentric circles ever have this condition? The area of the large circle minus the area of the small circle is the same as the area of the small circle? Draw the figure to be sure you understand the problem. Guess first and explain your guess, then solve for the answer.

Yes, when R equals r times the square root of 2, where R is the radius of the large circle and r the radius of the small circle.

Notes and comments

Keep a record of the problems missed and rework them in a week to two.

Chapter 11. Course Review, Grades 10-12
Section 11c. Decision Making Review

A review of the major points in **EVERYDAY DECISION MAKING.**

The major objective is for your student to understand that all decisions are based on undefined terms, defined terms, basic assumptions and previous decisions, laws or ordinances (called theorems in geometry). The following will recall and review the methods.

Methods for arriving at decisions range from guessing to formal logic. Everybody wants to make the correct decisions and they need to also know the weaknesses! Conclusions using various methods of decision-making are:

Illusions:
A method that uses what you think you see, which may not be the true or valid case. This type is used by witnesses to an event. This plays an important role in decision making and

reporting in the courts and even can change over a short time period or more so in years.

- **Guessing**
- **Listening to experts or non-experts on all sides of issues**
- **Observation**
- **Induction (predictions from past cases)**
- **Estimating**
- **Direct reasoning**
- **Indirect reasoning**
- **Statistics**
- **Data collection (random selection)**
- **Forms of an Implication (Converses, Inverses, Contrapositives)**
- **Ads (especially in creating a market)**

Comments and/or examples of the above are:

Observations

A method that uses what you think you see, which may not be the true case. This type is used by witnesses to an event. It plays an important role in decision making and reporting in the courts.

Examples: The following examples are two of the authors favorites. Do you see a young girl or an old lady or both in the drawing below? (Puck Magazine 1915 y W. Hill)

Another case:

Source Unknown

The teacher should try to adapt the problems to fit local experiences to make the problems more meaningful using the students' background, activities, interests, and future plans. The way a problem is stated can motivate the students to investigate it. The reason the model is Geometry was used is that it is a very simple deductive system at this stage, one that all can understand.

Guessing

This is a very weak way to make a decision, it amounts to basically tossing a coin to determine the decision. The only feature of it is you have passed the decision responsibility to something else such as a coin, but not the responsibilities resulting from the decision.

Inductive Reasoning (predictions)

This type arrives at a conclusion after investigating a few cases. This type requires record keeping to detect a pattern or trend. If the conclusion is a general one or about a future event, it may be invalid, since it is based on past events.

Example 1: How many chords can be determined by n points on a circle?

Draw a new circle for each case and count.

Points	Lines	
2	1	One chord through 2 points
3	3	Draw the figure.
4	?	Draw the case and what is your answer?

5	?	Draw the case and what is your answer? Do you see a pattern?
6	?	What is your answer? Do you see a pattern?
7	?	Answer for 7 points (Predict)
n	?	Formula?

Answer for 7 points is 21. N = P(P-1)/2
This is a prediction until proven.

Example 2: Repeat example 1, but this time instead of comparing points and chords compare points to regions the chords create within the circle.

Complete the following table.

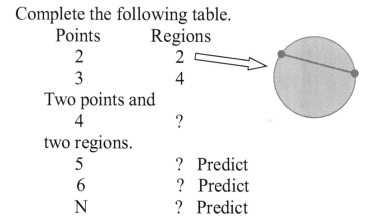

Points	Regions
2	2
3	4
Two points and 4	?
two regions.	
5	? Predict
6	? Predict
N	? Predict

Inductive Reasoning is drawing a conclusion as to future events from a few past events. This method for arriving at a conclusion is like predicting the future and is always questionable. The medical profession, weather forecasting, financial investors, auto repair, insurance companies, trades, many professions, and also our politicians use this type of reasoning to reach a conclusion. In example #2 you should have a contradiction for your prediction as the number for 6 points, which is a very good example of the potential weakness in inductive reasoning.

A different type of example of the induction weakness.

Example:
　A farmer feeds his turkey every morning for a month, but on Thanksgiving Day the farmer goes out to the feeding area and the turkey is in for a disappointment.
　What happened? (From a NCTM convention.)

Many times logical conclusions may not follow from the certain causes, since the information may be bias.

Example:
1. The junior class in a school A voted on the following question: Should cell phones be turned off in classes? (Only 15% voted no.) Which of the following statements printed in 5 different papers is the most correct?
    a. High School students vote to turn off cell phones in class.
    b. 85% of high school students vote to prohibit cell phones in school.
    c. Most students don't use cell phones in classes
    d. 15% of students use cell phones in classes
    Remember: Statements are true or false and conclusions are valid or invalid.

2. Fact: All students of school X, wear red caps at the football games. Fact: John is at the game and is wearing a red cap. Conclusions: Which are valid?
   a. John must be a student of school X.
   b. Olaf is not wearing a red cap at the game, then he is not a student at school X.
   c. Joe is not a student at X, then he does not wear a red cap at the game.

3. Listening to opinions by experts or non-experts. On TV or on radio does not make the person an expert, he or she could be a good reader, good looking, or may be well known such as an actor or athlete.

4. The problem with this method in #3 above is that issues are restated in various ways and often misinterpreted as to the truth of the statements and conclusions. What is the assumption #3 is based on?

Ads and Implications

5. The advertising world and other "worlds" (such as politics) use these to take advantage of people who do not understand them and the use of the forms of an implication.
    The following diagram will be used to explain the meaning of implications and their forms.

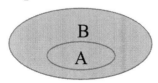

272

Look at the diagram and answer the following statements:

Original statement: If in A, then in B. (consider this valid)
Converse: If in B, then in A. (not always valid)
Inverse: If not in A, then not in B. (not always valid)
Contrapositive: If not in B, then not in A.(valid)

Implications are very important for interpretation of ads! Be sure your student understands them! Use a few ads from TV and what they hope you will interpret.

6. Fact: If you fly the Flag on Memorial Day, then you are patriotic. Assume valid, then which of the following are valid?
    a. If you are patriotic, then you fly the flag on Memorial Day. (Converse-may or may not be true or valid.)
    b. If you do not fly the flag on Memorial Day, then you are not patriotic. (Inverse-may or may not be true or valid.)
    c. If you are not patriotic, then you do not fly the flag on Memorial Day. (Contrapositive -True and valid.)

7. Fact: If you watch TV-PBS, then you want factual information. Assume the fact is valid, then which of the following may be valid?
    Converse (?)   Inverse (?)
    Contrapositive (T and valid)
   Hint: Draw the Implication figure (2 circles).

8. Fact: If you want factual information, then you watch TV-PBS. Hint: draw the diagram. Which of the following are valid or possibly valid if you assume the Fact is true and valid?
   a. If you don't watch TV-PBS, then you don't want factual information.
   b. If you don't want factual information, then you don't watch TV-PBS.
   c. If you watch TV-PBS, then you want factual information.

Answers
   a. Contrapositive - valid
   b. Inverse – false and possibly valid
   c. Converse – false and possibly valid

Data collection

Polls (Does your student understand what is meant by a poll?) are many times used to indicate a trend. Listed are questions you should know about polls. Give an example of each of the following and how it can influence the result.
   a. The number polled, reported statistically, plus where and how it was obtained.
   b. The selection process, time of day, the wording of the questions (randomness, age, location).
   c. How contacted: email, telephone, U.S. mail, or interview.

To be valid a poll the above should be explained.

Indirect Reasoning and Conclusion
A great case!

This is a very interesting indirect reasoning case. This type of reasoning, they say, is used much more than the direct reasoning to arrive at a conclusion. To review the indirect method: all the possible conclusions are listed and all but one is concluded as impossible, hence the remaining conclusion is correct. The following example will illustrate the method. Your student (in most cases that I have found) will have trouble with this. Take your time, even adults have trouble. Read it very carefully!

A CEO of a large company wanted to know which of his three advisers were the sharpest. He devised the following method to select the new chief adviser. He called them into his office, asked each one to go to a corner, and gave them the following instructions:
1. He will blindfold them and put either a White or Black beanie on each.
2. He will then remove the blind folds and if they see a black beanie they are to put their hand in above their head.
3. The first one to put his hand down and explain how he knew the color of his beanie would be my chief adviser. (He also added if the person was wrong that person would be fired immediately.)

The CEO then removed the blindfolds and each adviser put his hand up above his head. After a while one of the advisers put his hand down and proclaimed that he had a white Beanie on and explained how he knew. How did he know? Can you explain?

Hint: The key phrase is, "after a while," and then he said his beanie was white.

Here is the reasoning!

He knows his beanie is either white or black, but if his beanie were white, then the other two would know immediately their beanies are black. (Think about it.) But they did not put their hand down, so his hat had to be black!

Suggestion: Act it out and keep in mind he used indirect reasoning. (The author has used this in classes and it is good for the whole period with lots of participation.)

Again, conclusions are arrived at in everyday situations by using the following:

Suggestion: Discuss each of the following with your student, especially how they are misused.

- Guessing
- Listening to experts or non-experts on all sides of issues
- Observation
- Induction (predictions from past cases)
- Estimating
- Direct reasoning
- Indirect reasoning

- Statistics
- Data collection (random selection)
- Forms of an Implication
  (Converses, Inverses, Contrapositives)
- Ads (especially in creating a market)

Inductive reasoning example

Use your calculator to answer these questions.
1. Divide 1 by 9 and record your answer.
2. Divide 2 by 9 and record your answer.
3. Divide 3 by 9 and record your answer.
4. Divide 4 by 9 and record your answer.
5. Divide 5 by 9 and record your answer.
6. Divide 6 by 9 and record your answer.
7. Divide 7 by 9 and record your answer.
8. Divide 8 by 9 and record your answer.
9. Now write what you think is the answer when you divide 9 by 9?
   From the above your prediction is _____.
   Check your calculator's answer.

Note and comments

Index 1. Basic Postulates

These postulates are the ones for the author's geometry book. The set may be different for another book.

Postulate 1: A line has an infinite set of points.
Postulate 2: Two points will determine one and only straight line.
Postulate 3: There is a one to one correspondence between the points on a line and the real number system.
Postulate 4: The shortest distance between two points is a straight line.
Postulate 5: The length of a line segment is equal to the measure of the distance between the two points.
Postulate 6: The shortest distance between a point and a line on a plane is the measure of the perpendicular line segment. (Not valid in all geometries.)
Postulate 7: Three non-collinear point will determine a plane.
Postulate 8: If two parallel lines are crossed by another line (transversal) then the alternate interior angles are equal in measure.
Postulate 9: Through a point not on the given line, there is only one line through the point that is parallel to the given line (in plane geometry).
Postulate 10: The area of a rectangle is equal to the length times the width and the answer is in sq. units.: (Postulates11 and 12 are justified in chapter 8.)
Postulate 11: The formula for the circumference of a circle is $C = 2 \pi \Pi r$.
Postulate 12: The formula for the area of a circle is $A=\pi r^2$.

Postulate 13. The volume of a rectangular solid is the length times the width times the height. $V = lwh$.

Note: Postulates 11 and 12 are justified in the Trigonometry section and also listed as theorems in Index 3.

Index 2. Definitions

Definition 1: A Theorem is an important mathematical statement that follows logically from a set of undefined terms, definitions, postulates, or other theorems.

Definition 2: A postulate is a statement, which is assumed to be true.

Definition 3: The set of real numbers is the union of the rationals and the irrationals.

Definition 4: A rational number is a number, which can be written as the ratio of two integers.

Definition 5: An irrational number is a real number, which cannot be written as the ratio of two integers.

Definition 6: Inductive reasoning is arriving at a conclusion after observing a few cases. (The more cases, the more validity.)

Definition 7: Simple interest is the money paid for the use of the money borrowed. The formula for simple interest is $I = prt$.

Definition 8: Compound interest is interest not only paid on the loan but on the interest also. Formula for the total amount is $A = p(1 + r)^y$ (y is number of years). Note: Some Credit Cards companies use the formula, $A = p(1 + r/n)^{ny}$, where n is in months or days.

This is why credit card debt increases so rapidly and should never be allowed.
Example: Say you owe a $1000 at 7% interest (per year) and you pay it off after 2 months. What will you pay? $A = p(1 + r/12)^2 = 1000(1.00583)^2 = \$1011.69$

Definition 9: An ANNUITY is series of equal payments or deposits made for an agreed amount of time at a specified rate of interest.

Definition 10: A Permutation is the way a set of objects can be arranged where order **does** count. (Example: AB and BA are two permutations of AB.)

Definition 11: N!, read N factorial, means to multiply all the positive integers up to and including the number N.

Definition 12: A Combination is the way a set of objects can be arranged where order does **NOT** count. Example: AB and BA are one combination of AB.

Definition 13: Probability of an outcome is the ratio of the number of favorable outcomes or successes to (or divided by) the total number of possible outcomes.

Definition 14: Empirical Probability is the probability based on actual trials.

Definition 15: The Expectation of an event (e) is the probability of the event P(e) times the number of trials (T).
Formula: $E(e) = [P(n)](T)$.

Example: If you toss a coin 4 times, then how many heads would you expect? $E(H) = (1/2)4 = 2$

Definition 16: A game is Fair when the cost of playing equals the expected value times the winnings, or $n(c\$) = n(w\$)[P(w)]$,

> where:
> c$ is the cost per play.
> n is the number of plays.
> w$ is the payout for a win.
> P(w) is the probability of a win.

Example: It cost $2 to toss a coin and if a head turns up you win $4.
If you play 4 times, is it a fair game?
4(2)? 4(4)1/2
8 = 8   The game is fair, if the coin is fair!

Definition 17: The formula for odds in favor of an event is Odds = $P(e)/[1-P(e)]$.

Definition 18: An equation is a statement indicating two numbers are equal.

Definition 19: A formula is an equation representing a general rule such as a theorem.

Definition 20: A line is defined as the set of points that satisfy the equation $y = mx + b$, where x, y, and b are real numbers with the restriction that x and y both cannot be zero.

Definition 21: Direct Variation. If two variables are so related that y = kx, then x and y vary directly where k is the constant that relates the two variables. (This also indicates y/x = k.)

Definition 22: The SINE (SIN) of an acute angle in a right triangle is the ratio of the length of the side opposite the angle divided by the length of the hypotenuse.

Definition 23: The COSINE of an acute angle in a right triangle is the ratio of the length of the adjacent side divided by the length of the hypotenuse.

Definition 24: The TANGENT of an acute angle in a right triangle is the ratio of the length of the side opposite the angle divided by the length of the side adjacent to the angle.

Definition 25: Trig functions defined for the general angle. (The vertex of the angle is at the origin, one ray on the positive x-axis and the other ray is rotated counter clockwise. The "distance" in the following definitions refers to distance the point is from the origin.
  Sin A = y/R (opposite/hypotenuse or ordinate/distance)
  Cos A = x/R (adjacent/hypotenuse or abscissa/distance)
  Tan A = y/x (opposite/adjacent or ordinate/abscissa)

Definition 26: A Trig equation is an identity if the equation is true regardless of the value that is substituted for x or the variable

Definition 27: A Quadratic equation in two variables is defined as $y = Ax^2 + Bx + C$, where A, B and C are rational numbers and A is not equal to 0.

Definition 28: An equation is a function if the set of ordered pairs (x,y) satisfies the condition that for each x value there is only one y value. (There are other definitions.)

Definition 29: A parabola is the set of all points in a plane that are equal distance from a point (focus point) and a line called the directrix.

Definition 30: A circle is the set of points on a plane, which are equal distance from a given point called the center.

Definition 31: Inverse Variation results when two variables are related in such a way that their product is always the same number or constant. Symbolically, this is stated as $xy = k$.

Definition 32: The average speed is the total distance divided by the time it takes to travel the distance. $AS = D/T$

Definition 33: The MEAN or arithmetic average for a set of numbers is the sum of the numbers divided by n, the number of numbers in the set.
        Formula: Mean = (sum of n scores)/n

Definition 34: The MODE for a set of data is the most popular or most frequently occurring element in the set.

Definition 35: The MEDIAN for a set of data is the middle element when the elements are arranged in order of size or magnitude.

Definition 36: Properties of the Normal Curve:
    a. The mean, mode, and median are all equal or the same value and occur at the center or line of symmetry.
    b. One STANDARD DEVIATION on each side of the mean (line of symmetry) will include approximately 68% of the data.
    c. Two STANDARD DEVIATIONS on each side of the mean (line of symmetry) will include approximately 95% of the data.
    d. Three STANDARD DEVIATIONS on each side of the mean (line of symmetry) will include approximately 99.8% of the data.
    e. From the data or the curve, the RANGE can also be determined.

Definition 37: The RANGE is the difference between the highest or largest number and the lowest or smallest number in the set.

Definition 38: STANDARD DEVIATION is the square root of the mean of the squares of the deviations from the mean divided by n.

$$SD = \sqrt{\frac{(m-x)^2 + \ldots + (m-x_n)^2}{n}}$$

285
Definition 39: Properties of exponents
  a. x^m times x^n, where x is real and n and m are rational, then the product is x^{m+n}.
  b. x^m/x^n, where x is real and not equal to zero, and n and m are rational, then the quotient is x^{m-n}.
  c. x^0 is equal to 1 providing x is not zero.
  d. If x^{-n}, where x is real and not 0, and n is rational, then x^{-n} equals $1/x^n$.
  e. If $x^{n/d}$ and x is real, and n and d are rational and d is not zero then it is equal to:
  $$\sqrt[d]{x^n}$$
  f. If x is not zero, then the expression to $\sqrt[d]{x^n}$ = x to the n/d power and conversely. (see e, above.)
  g. If $(x^m)^n$ where m and n are rational numbers, then the product is x^{mn}.

Definition 40: If $n = b^L$, then the logarithm of n to the base b is L, and conversely. ($\log_b n = L$)

Definition 41: If Log n with no base indicated, then it means the base is 10.

Definition 42: Products by logs
    Log of (A times B) is the LogA + the LogB.
        (The bases must be identical.)

Definition 43: Quotients by logs
    Log of (A divided by B) is the LogA - the LogB.

Definition 44: Powers by logs
    Log of A^n is n times LogA.

Index 3. Essential Geometry Theorems

These theorems, it is assumed, were proven for and/or by the students in their geometry course and emphasized that all theorems (conclusions) are justified by undefined terms, defined terms, basic assumptions, and previous proven theorems (conclusions). Some converses are not listed!

Theorem 1: In a triangle the sum of any 2 sides is > then the third side and any side is greater than the absolute value of the difference of the other 2 sides. $A + B > C$ and $A > |B - C|$

Theorem 2: If 2 lines intersect, then the opposite angles are equal. (Some books call these vertical angles.

Theorem 3: The sum of the angles in a triangle is 180 degrees.

Theorem 4: The exterior angle of a triangle is equal to the sum of the 2 non-adjacent angles.

Theorem 5: If 2 parallel lines are crossed by another line (transversal), then the alternate interior angles are equal.

Theorem 6: If 2 lines are crossed by another line and the alternate interior angles are equal then the 2 lines are parallel. (Indirect proof)

Theorem 7: Two triangles are similar (~) if:
    a. two angles are equal in each triangle. (AA)
    b. sides in each triangle are in equal ratio. (SSS)

    c. 2 sides (equal ratio) and the include angles (equal) in each. (SAS)
    d. 2 equal angles and the included sides in equal ratio. (ASA)

Theorem 8: In an Isosceles triangle the angles opposite the equal sides are equal in measure.

Theorem 9: The angles in an equilateral triangle are equal in measure.

Theorem 10: In a scalene triangle, the larger the side the larger the opposite angle.

Theorem 11: In a parallelogram:
    a. the diagonals bisect each other.
    b. opposite angles are equal.
    c. the opposite sides are equal.

Theorem 12: In a rectangle:
    a. diagonals are equal in measure.
    b. all the properties of a parallelogram.

Theorem 13: If the diagonals of a parallelogram are equal, then the figure is a rectangle.

Theorem 14: In a square:
    a. diagonals are perpendicular.
    b. all the properties of a rectangle are valid.

Theorem 15: The area of a triangle is ½ the base times the height. $A = (½) bh$. Area is in square units.

Theorem 16: The area of a parallelogram (Measurements in the same units):
    a. Parallelogram is base times height. A = bh.
    b. Trapezoid is the sum of the bases times ½ the height. A = (1/2)h(B+b)

Theorem 17: Any point on the perpendicular bisector of a line segment is equal distance form the end points of the segment.

Theorem 18: In a right triangle with sides a, b, and c, the hypotenuse, then $a^2 + b^2 = c^2$. Pythagorean Theorem (VIT very important theorem)

Theorem 19: In a 30-60 right triangle the side opposite the 30 degree angle is ½ the hypotenuse and the other side is (h/2)√3 where h is the hypotenuse.

Theorem 20: The three altitudes of a triangle are concurrent or intersect at one point. (The point may be outside the triangle.)

Theorem 21: The three medians of a triangle are concurrent at a point 2/3 the length of each median.

Theorem 22: The three angle bisectors of a triangle are concurrent at a point equal distance from the sides.

Theorem 23: If a point is on the angle bisector, then the point is equal distance from the sides of the angle.

Theorem 24: The segment from the center of a circle to the point of tangency is perpendicular to the tangent.

Theorem 25: Given three points, the center of the circle (O), point of tangency (T), and any other point on the tangent (B), then segments OT squared + BT squared = OB squared. ($a^2 + b^2 = c^2$)

Theorem 26: The perpendicular bisectors of the chords in a circle intersect at the center of the circle.

Theorem 27: An inscribed angle in a circle is equal to ½ the degree measure of the intercepted arc.

Theorem 28: In a plane figure, if the dimensions are doubled, then the area is 4 times the original area.

Theorem 29: If a line is perpendicular to 2 lines in a plane at the point of intersection, then it is perpendicular to the plane. (The telephone pole theorem.)

Theorem 30: The volume of a prism or cylinder is the area of the base times the altitude (height). $V = Bh$ answer in cubic units.

Theorem 31: The volume of a pyramid is 1/3 the area of the base times the height. $V = (1/3)Bh$ cubic units

Theorem 32: The volume of a cone is 1/3 the area of the base times the height: $V = (1/3)\pi r^2 h$ cubic units

Theorem 33: The lateral area of a cone is:
L = ½ Ch sq. units   or L = (1/2)2πrh or L= πrh
Where C = circumference, r = radius, h = altitude

Theorem 34: The volume for a sphere is V = (4/3) πr³ cu. Units

Theorem 35: The area of a sphere is A = 4 πr² sq. units.

Theorem 36: The ratio of the areas of 2 spheres is a/A = (r/R)².

Theorem 37: Theorem 36. The ratio of the volumes of 2 spheres is a/A = (r/R)³.

Trigonometry

Sin Theorem 38: If the sides of a triangle ABC are a, b, and c, then sinA/a = sinB/b = sinC/c.

Cosine Theorem 39: If the sides of a triangle are a, b, and c, then:

$$c^2 = a^2 + b^2 - 2ab(\cos C)$$

or

$$a^2 = b^2 + c^2 - 2bc(\cos A)$$

or

$$b^2 = a^2 + c^2 - 2ac(\cos B)$$

Theorem 40: Given any angle, then sin²A + cos²A=1

Theorem 41: If A is an angle, then Tan A = sinA/cosA for values of A except for odd multiples of 90 degrees.

Algebra

Theorem 42: A formula for the value of Pi is the limit of;
$$Pi = n(\sin 180/n) \text{ as n get very large.}$$

Theorem 43: If given the quadratic equation, $Ax^2 + Bx + C = 0$ then the
solution is:
$$x = \frac{-B \pm \sqrt{B^2 - 4AC}}{2A}$$
(At this level of math $B^2 - 4AC > 0$)
  a. The sum of the answers equals $-B/A$.
     Proven in Chapter 8.
  b. The product of the answers equals C/A.
     The product is left for the student to prove.
     (Easy!)

Theorem 44: Given the quadratic equation, $y = Ax^2 + Bx + C$, then the turning point on the graph has the coordinates:
$$\{-B/2A, f(-B/2A)\}.$$
The equation for the line of symmetry is $x = 2A$.

Theorem 45: If given an inequality, then the inequality can be multiplied(divided) by a negative number and the result is an inequality of the opposite order.

Theorem 46: If two numbers have a common divisor, then their difference (a-b) is divisible by the common divisor.

Theorem 47: If given an inequality, then the inequality can be multiplied (divided) by a negative number and the result is an inequality of the opposite order.

Theorem 48: Jordan Curve Theorem (not proved): Two points, A and B in a maze are on the same side if:
    a. The segment connecting A to B crosses an even number of segments.
    b. The two points are on opposite sides if the crossing is an odd number of segments.

(A maze can be traversed by keeping the right or left hand on the wall and follow the wall. It may not be the shortest path but you will arrive at the exit.)

LOGARITHMS

Theorem 49: If $Log_b N = X$, then X equals = logN/Logb. Example: $12 = 3^n$, then log12/log3 = n

NUMBER THEORY (not justified)

Theorem 50: If two numbers have a common divisor, then their difference (a-b) is divisible by the common divisor.

Theorem 51: The sum of the first n odd integers is n^2.

Theorem 52: The sum of the first n even integers is n(n+1).

Theorem 53: The sum of the first n integers is n(n+1)/2.

Theorem 54: If the digits of a number add to 9, then the number is divisible by 9.

Index 4. Quotes

Geometry without rigor is not Geometry, but Geometry with too much rigor only develops rigor mortis.

Frank Allen
Former President of NCTM

Mathematics is not a careful march down a well-cleared Highway, but a journey into a strange wilderness, where explorers often get lost.

W. S. Anglin

Neglect of mathematics works injury to all knowledge.

Roger Bacon

Mathematics is the gate and the key to all sciences. He who is ignorant of it cannot know the things of this world.

Roger Bacon

Mathematics, the unshaken Foundation of Sciences, and the plentiful Fountain of Advantage to human affairs.

Issac Barrow

Young Cayley developed an amazing skill in long numerical calculations.

E.T. Bell

Alexander the Great asked his teacher if there was an easier way to learn Mathematics. The teacher replied: "There is no royal road to Geometry."

E.T. Bell

Students of mathematics ... the first time something new is studied seem they hopelessly confused. Then, upon returning, (to the concept) after a rest, everything has fallen into place.

E. T. Bell

C Hipparchus of Nicaea, (180-125 B.C.E.) compiled the first trigonometric table.

C.B. Boyer

History shows that those heads of empires who have encouraged the cultivation of mathematics, are also those whose reigns have been the most brilliant and whose glory is the most durable.

Michel Chasles

The miraculous powers of modern calculations are due to three (mathematical) inventions: Hindu Notation, Decimal Fractions, and Logarithms. (What would you add today?)

F. Cajorie

The invention of logarithms Laplace said amounted to "shortening the labours (and) doubled the life of the astronomer."

F. Cajorie

Mathematics seems to endow one with something like a new sense.

Charles Darwin

I think therefore I am.

                                                          Rene Descartes

Factual science may collect statistics and make charts. But its predictions are, as has been well said, but the Past is history reversed

                                                          John Dewey

There are certain branches of mathematics where calculation conserves its rights.

                                                         P.G.L. Dirichlet

It is easier to square the circle then get around a mathematician.

                                                        A. De Morgan

The important thing is to not stop questioning.

                                                      Albert Einstein

Factual science may collect statistics and make charts. That they (all citizens) might excel in public discussions on philosophic or scientific questions, they must be educated (rhetoric, philosophy, mathematics, and astronomy).
                The Athenian Sophist School Curriculum (480 B.C.E.)
                                                            F. Cajori

Let no person (high school student) ignorant of Geometry exit here.

                                                           J. Elander

It is not how much you cover, but how much you uncover.
H. Fawcett

All numbers in the form of 4n+1 are the sum of 2 squares.
P. Fermat

There cannot be a language (mathematics) more universal...and more worthy to express the invariable relations of the natural things.
Joseph Fourier

Mathematics through the power of computers pervades almost every aspect of our lives.
David L. Goines

Mathematics has been a human activity for thousands of years. To some small extent everybody is a mathematician and does mathematics consciously.
P. Davis & R. Hersh

In the ancient world as now, trade has been the principal consumer of mathematical operations.
P. Davis & R. Hersh

The heart of the mathematical experience is, of course, mathematics itself.
P. Davis & R. Hersh

Mathematics is about anything as long as it is a subject that exhibits the pattern of assumption-deduction-conclusion.

            P. Davis and R. Hersh

The theory of probability entered mathematics through gambling.

             P. Davis & R. Hersh

The Great Architect of the Universe now begins to appear as a pure mathematician.

                J. H. Jeans

The geometric properties of the conics were worked out in a thorough, beautiful, but not simple algebraic form by Appolonius.

              Edna E. Kramer

GOD gave us the integers (whole numbers) and all the rest is the work of man.

               L. Kronecker

You cannot fake. In mathematics, no one can be fooled. You can either prove or you cannot.

               Jerry P. King

A mathematician, like everyone else, lives in the real world. But the objects with which he works do not. They live in that other place--the mathematical world. Something else lives here also. It is called TRUTH.

               Jerry P. King

Attributing teaching and learning failure to something called "math anxiety" serves no purpose except to provide a built-in excuse for inadequate performance on both sides.
>
> Jerry P. King

The essence of plane analytic geometry lies in the matching of ordered pairs of real numbers with the points of a plane.
>
> Edna E. Kramer

To measure is to know.
>
> Johann Kepler

To measure the unmeasurable.
>
> Johann Kepler

In truth, all of life, in one way or another is concerned with the study probability.
>
> H. Gross & F. Miller

In short, the house plays the percentages, while the player relies on luck.
>
> H. Gross & F. Miller

Analytic Geometry...constitutes the greatest single step ever made in the progress of the exact sciences.
>
> John Stuart Mill

YOUNG PEOPLE WHO HAVE ACQUIRED THE ABILITY TO ANALYZE PROBLEMS, GATHER INFORMATION, PUT THE PIECES TOGETHER

TO FORM TENTATIVE SOLUTIONS WILL ALWAYS BE IN DEMAND.

> J. G. Maisonrouge
> Board Chairman
> IBM World Trade Corp.

The advance and the perfecting of mathematics are closely joined to the prosperity of a nation.

> Napoleon

Let no man ignorant of Geometry enter here.

> Plato (±400 BC)

Number rules the universe.

> The Pythagoreans (±550 B.C.)

The connection between the improvement of human conditions and the happiness of the human race is Science. (The Queen of the sciences is MATHEMATICS.)

> Neil Postman

Just as statistics has spawned a huge testing industry, it has done the same for the polling of "public opinion."

> Neil Postman

A new technology does not add or subtract something. It changes everything.

> Neil Postman

Pythagoras, the teacher, paid his student three oboli (a coin) for each lesson he attended and noticed that as the weeks

passed the boy's initial reluctance to learn was transformed into enthusiasm for knowledge. To test his pupil, Pythagoras pretended that he could no longer afford to pay the student and that the lessons would have to stop, at which point the boy offered to pay for his education.
Simon Singh

Relationships between different subjects (even branches of mathematics) are creatively important in mathematics.
Simon Singh

Mathematics consists of islands of knowledge in a sea of ignorance.
Simon Singh

It can be of no practical use to know that π is irrational, but if we can know, it surely would be intolerable not to know.
E. C. Titchmarsh

Statistical thinking will one day be as necessary for efficient citizenship as the ability to read and write.
H. G. Wells

The definition of a good mathematical problem is the mathematics it generates rather than the problem itself.
Andrew Wiles

Many of the laws of the sciences are stated in the language of variation.
Unknown

Thinkers recognize when two variables are related, but it is Mathematics that connect them numerically.

Unknown

"Decision making" is easy to read but how do you do it!

Unknown

Mathematics is like a mighty tree with number (counting numbers) for its roots.   Arithmetic grows on numbers, algebra on arithmetic, algebra, and geometry. Calculus builds on all four. It is a tree that grows in time, fertilized by the minds of mathematicians and the applied needs of society.

Unknown

We learn the new in the light of the old.

Unknown

The proof of the pie is in the eating.

Unknown

To Think is to Know.

Unknown

Statements are true or false. Conclusions are valid or invalid.

Unknown

Mathematics is not a spectator sport!

Unknown

Understanding evolves from work, appreciation from applications.

Unknown

Index 5. Conversion Table

Length

English		Metric or SI
1 inch (in.)	=	2.54 centimeters (cm.)
.39 in.	=	1 cm = .1 decimeter (dm.)
12 in. = 1 foot (ft.)	=	30.48 cm.
3 ft. = 1 yard (yd.)	=	.9144 m.
39.37 in.	=	1 meter (m.) or 100 (cm.)
5280 ft. = 1 mile(mi)	=	1.609 kilometers (km.)
.621 mile	=	1 km.

Area

1 sq. ft. = 144 sq. in. = = .092903 sq. meters
1 sq. yard = 9 sq. ft. = .836 sq. meters
1.196 sq. yd. = 10.765 sq. ft. = 1 sq. meter
1 sq. miles = 640 acres = 259.004 hectares
2.471 acres = 1 hectare
1 acre = 43560 sq. ft. = .405 hectare

Volume

1 cubic in. = 16.388 cu. Cm.
1 cubic ft = 1728 cu in. = 28318.46 cu cm
1 cu. yd. = 27 cu. ft. = .765 cu meters
1.308 cu. yds. = 35.31 cu. ft. = 1 cu. meter
1 U.S. Gal. = 4 qts. = 8 pts. = 128 fluid oz. = 231 cu in.
*1 U.S. gal = 3.785 liters
*1 British gal = 4 liters = 4000 cu. cm.=1.0566 U.S. Gal
(This is why a gallon of gas in Canada cost more than a U.S. gallon.)

A few kitchen conversions

1 teaspoon = 5 milliliters (ml)
1 tablespoon = 15 ml
1 oz. = 29.6 ml

Printed in Poland
by Amazon Fulfillment
Poland Sp. z o.o., Wrocław